URBANISING BRITAIN

URBANISING BRITAIN brings together the work of some of the leading British historical geographers of the younger generation to consider nineteenth-century urbanisation as a process, emphasising the dimensions of class and community. The essays in this collection reflect the increasing use of social science concepts within the field of historical geography and cover various aspects of urbanisation from its origins in migration to its consequences in urban culture and public health. The contributions combine conceptual sophistication with original empirical research to present a series of important and innovative statements about the changing nature of the Victorian city, and reflect the value of a critical theoretical perspective, hitherto absent from much work in this area.

T0283700

Cambridge Studies in Historical Geography

Series editors:
ALAN R. H. BAKER/J. B. HARLEY/DAVID WARD

Cambridge Studies in Historical Geography encourages exploration of the philosophies, methodologies and techniques of historical geography and publishes the results of new research within all branches of the subject. It endeavours to secure the marriage of traditional scholarship with innovative approaches to problems and to sources, aiming in this way to provide a focus for the discipline and to contribute towards its development. The series is an international forum for publication in historical geography which also promotes contact with workers in cognate disciplines.

1 Period and place: research methods in historical geography. *Edited by* A. R. H. BAKER *and* M. BILLINGE
2 The historical geography of Scotland since 1707: geographical aspects of modernisation. DAVID TURNOCK
3 Historical understanding in geography: an idealist approach. LEONARD GUELKE
4 English industrial cities of the nineteenth century: a social geography. R. J. DENNIS*
5 Explorations in historical geography: interpretative essays. *Edited by* A. R. H. BAKER *and* DEREK GREGORY
6 The tithe surveys of England and Wales. R. J. P. KAIN *and* H. C. PRINCE
7 Human territoriality: its theory and history. ROBERT DAVID SACK
8 The West Indies: patterns of development, culture and environmental change since 1492. DAVID WATTS*
9 The iconography of landscape: essays in the symbolic representation, design and use of past environments. *Edited by* DENIS COSGROVE *and* STEPHEN DANIELS*
10 Urban historical geography: recent progress in Britain and Germany. *Edited by* DIETRICH DENECKE *and* GARETH SHAW
11 An historical geography of modern Australia: the restive fringe. J. M. POWELL*
12 The sugar-cane industry: an historical geography from its origins to 1914. J. H. GALLOWAY
13 Poverty, ethnicity and the American city, 1840–1925: changing conceptions of the slum and ghetto. DAVID WARD*
14 Peasants, politicians and producers: the organisation of agriculture in France since 1918. M. C. CLEARY
15 The underdraining of farmland in England during the nineteenth century. A. D. M. PHILLIPS
16 Migration in Colonial Spanish America. *Edited by* DAVID ROBINSON
17 Urbanising Britain: essays on class and community in the nineteenth century. *Edited by* GERRY KEARNS *and* CHARLES W. J. WITHERS

Titles marked with an asterisk * are available in paperback

URBANISING BRITAIN

Essays on class and community
in the nineteenth century

Edited by
GERRY KEARNS
and
CHARLES W. J. WITHERS

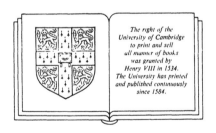

The right of the
University of Cambridge
to print and sell
all manner of books
was granted by
Henry VIII in 1534.
The University has printed
and published continuously
since 1584.

CAMBRIDGE UNIVERSITY PRESS

CAMBRIDGE
NEW YORK   PORT CHESTER
MELBOURNE   SYDNEY

CAMBRIDGE UNIVERSITY PRESS
Cambridge, New York, Melbourne, Madrid, Cape Town, Singapore, São Paulo

Cambridge University Press
The Edinburgh Building, Cambridge CB2 8RU, UK

Published in the United States of America by Cambridge University Press, New York

www.cambridge.org
Information on this title: www.cambridge.org/9780521364997

First published 1991
This digitally printed version 2007

A catalogue record for this publication is available from the British Library

Library of Congress Cataloguing in Publication data
Urbanising Britain : essays on class and community in the nineteenth
    century/edited by Gerry Kearns and Charles W. J. Withers.
        p.   cm.–(Cambridge studies in historical geography 17).
    Includes bibliographical references and index.
    ISBN 0 521 36499 X
    1. Urbanization–Great Britain–History–19th century.
    2. Community development. Urban–Great Britain–History–19th
    century. 3. Social classes–Great Britain–History–19th century.
    I. Kearns. Gerry. II. Withers, Charles W. J. III. Series.
HT384.G7U73 1991
307.76'0941'09034–dc20 90–2558   CIP

ISBN 978-0-521-36499-7 hardback
ISBN 978-0-521-04609-1 paperback

Contents

List of figures   page   vi
List of tables   vii
Preface   viii
Notes on contributors   ix

INTRODUCTION: 1
Class, community and the processes of urbanisation
GERRY KEARNS and CHARLES W. J. WITHERS

1   BIOLOGY, CLASS AND THE URBAN PENALTY   12
    GERRY KEARNS

2   PUBLIC SPACE AND LOCAL COMMUNITIES: 31
    The example of Birmingham, 1840–1880
    BILL BRAMWELL

3   CLASS, CULTURE AND MIGRANT IDENTITY: 55
    Gaelic Highlanders in urban Scotland
    CHARLES W. J. WITHERS

4   THE COUNTRY AND THE CITY: 80
    Sexuality and social class in Victorian Scotland
    J. A. D. BLAIKIE

5   MOBILITY, THE ARTISAN COMMUNITY AND POPULAR
    POLITICS IN EARLY NINETEENTH-CENTURY ENGLAND   103
    HUMPHREY SOUTHALL

Notes   131
Consolidated bibliography   154
Index   172

Figures

1.1   Age-specific mortality for social groups (males only) in   page 16
      the urban and rural registration divisions of England and
      Wales, 1871

1.2   Causes of death for occupied males aged 25–64 in the   16
      different social groups 1900–1902

1.3   The relative mortality of urban and rural areas for males   17
      and females, England and Wales, 1851–1860 and
      1891–1900

1.4   Mortality for males aged 25–44 in different social groups   18
      at selected dates, 1860–1902

2.1   A courtyard with courtyard housing and communal   38
      facilities, off Allison Street, Birmingham, c. 1871

2.2   Distribution of municipal parks in Birmingham,   50
      1840–1880

5.1   Movements of Thomas Watson, 1839–1842   108

5.2   Movements of Ellis Rowland, 1835–1846   110

5.3   Movements of Robert Gammage, 1840–1853   111

5.4   Movements of Feargus O'Connor, 1838 and 1839   125

Tables

3.1 Birthplace, by county, of the Highland-born population   page 63
of selected Lowland towns, 1851

3.2 Gaelic speaking in urban Lowland Scotland, 1891: the   65
examples of Aberdeen, Dundee, Perth and Stirling

3.3 Residential mobility and length of residence in Glasgow   72
of a sample of Highland-born residents in City Parish,
Glasgow, 1851–1897

5.1 Mobility between 1835 and 1846, by date of birth   115

5.2 Levels of tramping by age group   115

Preface

The authors of these essays were all students at the Department of Geography, Cambridge. We have all completed theses in historical geography at various places and on the basis of our common substantive and theoretical interests decided to explore the possibilities of working together. We have met on a number of occasions over the past four years to discuss general issues of interest as well as to present and debate the essays in this book. We would all like to acknowledge the benefits we have gained from this collective work and would recommend such a mutually supportive pattern of work to others. We would like to acknowledge the help and support we have received from Dave Green, who, although he is not represented in this volume, was an integral member of our discussion group. We would also like to thank Derek Gregory, without whose teaching we would certainly not be interested in the topics of this volume in quite the same way.

Notes on contributors

ANDREW BLAIKIE is Lecturer in Gerontology, Centre for Extra-Mural Studies, Birkbeck College, University of London. He is preparing a book on illegitimacy in nineteenth-century northeast Scotland to be published by Oxford University Press. His research interests include the social history of ageing and he is currently Director of the M.Sc. programme in Life Course Development at Birkbeck.

BILL BRAMWELL (MTS, MILAM, MA, PH.D.), after a period of research in historical geography at London University and teaching at the West Sussex Institute of Higher Education, joined the English Tourist Board as Area Development Manager. He was recently appointed Senior Lecturer in Tourism at Sheffield City Polytechnic.

GERRY KEARNS is a Lecturer in Geography in the Department of Geography at the University of Liverpool. He is currently researching and teaching the history of urban public health. His other research interests include the history and philosophy of geography.

HUMPHREY SOUTHALL is a Lecturer in Geography at Queen Mary and Westfield College, University of London. His interest in nineteenth-century artisans and their organisation developed out of research on unemployment and labour migration using trade union records. He is currently directing a research project on the origins of the north–south divide in Britain.

CHARLES WITHERS is Head of the Department of Geography and Geology at Cheltenham and Gloucester College of Higher Education. He has published on the historical geography of Gaelic Scotland and on Highland–Lowland migration. He is co-editor (with Aidan McQuillan) of the Historical Geography Research Group Research Paper Series. In addition to further work on urban migrant culture, his current research interests include the culture of scientific enterprise, and Scottish mobility patterns in European context.

Introduction: class, community and the processes of urbanisation

GERRY KEARNS AND CHARLES W. J. WITHERS

During the nineteenth century Britain became both an urban and an industrial society. The census of 1851 reported significantly that over half the population of England and Wales lived in towns. By 1801, one-third of the English population could be classified as urban, but only one in ten town dwellers lived in the big cities of 100,000 people or more. By 1901, three-quarters of the English population was urban, of whom about one-third lived in big cities.[1] The industrialisation of the nineteenth century was more exclusively urban and happened far more quickly than hitherto. The earliest decades of industrialisation in the new cities were particularly based on the recruitment of labour from beyond the city. The cities effectively pulled free from earlier seasonal and circular patterns of migration which had balanced the sometimes competing and more often complementary demands of town and country. British cities, though, continued to be fed primarily by internal migration and offered nothing comparable to the melting-pots of North America or Eastern Europe.[2]

The dramatic recasting of the spatial structure of British society has attracted the attention of sociologists, economists, historians, anthropologists and geographers, as it attracted, too, the attention of contemporaries. Some description of this transformation is present in almost all general accounts of nineteenth-century Britain. The processes of urbanisation and their consequences have also been studied in detail in more specialised monographs.[3] In Britain, historical geographers may, as Dennis and Prince suggest, have paid far too little attention to towns in the past, looking instead at rural landscapes.[4] The belated growth of British urban historical geography in the 1970s and 1980s was especially marked by an emphasis on the patterns of residential segregation within cities and, although similar concerns were shown in the work of American urban historians such as Warner, Thernstrom, Hershberg and Zunz, American historical geographers remained more concerned with regional cultural geography in the tradition of Carl Sauer, an approach much closer to the work of Darby and his

1

collaborators than to the concerns either of British urban historical geographers or American urban historians.[5]

The essays in this book cover many of the topics treated by earlier historical geographers. In studying migration and its consequences, Blaikie, Withers and Southall are drawing on earlier studies such as those by Darby, Smith, Johnson, Lawton and Pooley.[6] In looking at the way work, home and community were related, Bramwell has taken up a theme previously studied by Vance.[7] Geographers such as Howe and Gilbert have looked at disease in the city and this is the concern of the paper by Kearns.[8] The intellectual context in which we are working, though, is continually changing. Migration is not taken up here in quite the way that the spatial science tradition reaching back to Ravenstein has treated it. The relations between work, home and community are not covered by Bramwell in the same manner as they were by Vance. Kearns is not retracing the epidemiological perspective of Howe and Gilbert. These differences partly reflect the unpredictable play of intellectual curiosity, but they are also shaped by broad changes in the concerns of historical geographers. It is the purpose of this introduction to sketch some of the features of this changing scene.

Urban ecology and historical geography

For many writers the spatial rearrangement of social and economic life was and is a symptom of more fundamental changes in society and economy. The new spatial structure is principally treated as posing problems of adjustment for society: the *anomie* of the uprooted, the risk of moral and biological contagion, the challenge of housing the poor, the necessity for intra-urban transportation, the problem of public order and so on. These topics assume greater or lesser significance in historical writing relative, at least in part, to broader contemporary political and social concerns. This is clear in the approach of the urban ecologists. Two things characterise this work: an emphasis on broadly biological processes of city growth and social adaptation and an interest in the logistics of social interaction and cohesion. The first covers the well-known influence of Darwinian plant ecology on the Chicago school of urban sociology.[9] The second is quite explicit in the writings of Simmel and Park and has been developed further by others.[10] The biological perspective was always under question in sociology. The worries of Max Weber, writing contemporaneously with Park and Simmel, about the applicability of these natural science concepts to the social sciences are only the best known of a set of writings dating from what Hughes has termed the attack on positivism of the 1890s.[11] Furthermore, in sociology, spatial relations as such received less attention than the social organisation of groups.[12] Yet questions of social cohesion in the face of ethnic segregation (which, in part, animated the Chicago school of the 1920s) did not evaporate. They were repeatedly put back on the agenda by black civil rights groups. Measurement and discussion of segregation has thus remained a crucial and active issue within urban sociology.[13] It was this reduced version of a purely quantitative urban ecology that attracted the urban geographers of the 1960s and 1970s.[14] Historical data sets were often used to illustrate the testing of these models.[15] Factor analysis and associated correlation techniques corresponded to a 'stripped-down' version of urban sociology whose aim was to describe urban structure with reference to a small group of spatial models. The existence of the 'natural areas' on which the exercise depended might be demonstrated by showing the range of economic, cultural and attitudinal variables that could be mapped to reveal essentially the same underlying spatial structure. It is only in this sense that Robson could claim in 1969 that 'increasingly, human geographers are including aspects of sociological material within their field of interest'.[16]

It would have been possible to show that the spatial science perspective of the geographers had offered them a blinkered view of the intellectual range of the early urban ecologists and, in their different ways, the discussions of Park and Simmel in the recent works of Ward and Harvey show this very clearly.[17] Further, the intellectual agenda of urban historical geography could have been regenerated by a more careful reconsideration of the social and political theories explicit in the work of the Chicago school. Instead, historical geographers looking at the city have responded to changes in urban and social history and changes in human geography.

The first of these pressures for change came from urban historical geographers' greater contact with urban history and, in particular, from the flowering of urban social history. Studies such as Pritchard's on Leicester incorporated an analysis of the building process as a producer of the urban spatial structure. Following the work of Dyos on Camberwell and Beresford on Leeds, there was a series of works in urban history that looked at the policies of landowners and the effect of the pre-urban cadastre on city-form.[18] This was taken up by several geographers.[19] There were also monographs in social history, influenced to a greater or lesser extent by Marxism, which took particular cities as containers for significant social processes. The question of the place of an aristocracy of labour in defusing revolutionary pressure in Britain, which goes back explicitly to Lenin, was discussed in general terms by Hobsbawm and was taken up in local case studies by Gray and Crossick and, most excitingly, in a comparative work by Foster.[20] Hobsbawm's work has been tremendously important in integrating social and economic issues and in framing some of the 'big questions' that dominate evaluations of the industrial revolution. Edward Thompson's studies of the changing social relations of production incorporated cultural and political alongside economic factors.[21] Thompson's moral and political commitment held both materialist and empathetic approaches in a creative tension that continues to inspire even while its theoretical implications are still being evaluated.[22] The sociology of social stability concerned not only labour historians but also scholars working on disease and poverty.[23] Stedman Jones' promiscuous sampling of sociological theory as well as the wide-ranging nature of his synthesis has not been equalled and *Outcast London* has done much to extend the research agenda of urban social history and subsequently urban historical geography. There were simply too few urban historical geographers to take up all the topics raised in urban history, from politics to leisure, but there was at least a familiarity with the range of these interests and a sense that if geographical accounts of the city did not take some of these on board, then they might end up being far too narrow.[24] Geographers were certainly in debate with urban historians both at conferences and in print.[25] In the mid-1970s Cannadine, an urban historian, and Ward, an historical geographer, debated the date at which British cities could be described as 'modern' simply in terms of Burgess' original model of transport development and the emergence of rings of distinct social status.[26] When, in 1982, Cannadine urged that the debate should shift from the explanation of urban form to the study of the interrelations between spatial structures on one hand and social and political processes on the other, he was doing no more than echoing developments within human geography itself.[27]

The best insulation against the sprawling concerns of social history might, quite reasonably, have been to suggest that, interesting as all this was, it was not geography. However, with their own questioning of the value and methods of the spatial–science approach, human geographers were becoming highly eclectic.[28] The developments in geography had two related features: first, the advocacy of Marxist approaches and, secondly, the more general critique of positivism. In 1973 Brookfield and Harvey set out Marxist perspectives for, respectively, development and urban studies in geography.[29] These two surveys were very critical of the current models based on neoclassical economics and offered instead the politically committed project of historical materialism. Human geography began to be reintegrated into social science, no longer with just one possibly outdated school but with the whole debate about the grounds of knowledge in social science and the consideration of Marxist and non-Marxist alternatives.[30] These tendencies were reinforced by an accompanying critique of positivism. From a rejection of bogus claims to value-neutrality in human geography, this moved to a wholesale consideration of the epistemology of the social sciences. Gregory claimed for human geography a place among the social sciences and urged that geographers contribute to Habermas' project of a critical social science.[31] Gregory argued that geography was a social science and that historical questions were central to many of the really exciting controversies in the social sciences.

These challenges cannot be met without a fundamental rethinking of the urban ecological tradition in historical geography. Such a re-evaluation is implicit in the gap between the best of current research and the lifeless rings and sectors of earlier studies. Once the theory of natural areas is abandoned, then the endless correlation of distributions based on censuses, rate books and directories and covering ever more imaginative measures of residential persistence and class composition becomes less and less defensible. If the 'natural areas' of the urban ecologists do not exist, then we either abandon the pretence of a separate spatial perspective or we urge it in rather different terms. Only with urban ecology can the spatial pattern be both explanaans and explanandum.

In this respect there is some justice in Pooley's harsh judgement on Dennis' comprehensive survey of urban historical geography: it came too late.[32] It essentially reported on Ph.D. research of the early 1970s, by himself and others, but it took on board what Baker has called the 'conceptual revolution' of late-seventies' human geography. Dennis' shotgun marriage of urban ecological research with anti-positivist and philosophical critique produces a curious diffidence between the textbook sheets. If Dennis' book was too late, was it not also too early? It is a helpful oversimplification to say that theory ran ahead of research in human geography. If historical questions really are basic to many of the contentious issues in social science, then there must be some dialogue between theoretical inquiry and empirical research. The theoretical ferment of the last decade in human geography has emphasised the inevitable interplay between theoretical and empirical concerns. There are a number of tasks to be undertaken before any summary view of nineteenth-century British urbanisation, such as that attempted by Dennis, could be offered with any pretence of integrating empirical studies into the broader intellectual agenda to which many geographers are now committed. Not least among these, of course, is the need to continue with new research. The toing and froing between generalisation and research cannot be bypassed through the use of ideal types. Only an intimate familiarity with the variety, scale and contemporary opinions revealed by making and reading detailed case studies can guard the geographer, as well as the historian, against the dangers of ahistorical and hasty generalisation. At the same time, careful attention to the theoretical and empirical basis of existing generalisations is necessary before they are used, as they must be, both to frame case studies and then to evaluate the significance of the findings. Neither the insensitivity of the grand theorist anaesthetised against the irritation of contextual variety nor the blithe disregard of the antiquarian wallowing in the celebration of local specificity will answer the challenge thrown down by the final death throes of urban ecology.

These essays have tried to handle that tension creatively. In no sense do they add up to a systematic re-examination of nineteenth-century urbanisation considered from a geographical point of view. Questions of ideology and culture have perhaps received most attention and the description of the economic structure is probably not much advanced by these essays. Like all such collections, there are differences of emphasis and approach among these authors, but their common influences do, we believe, show through. The rest of this introduction sketches some of that common ground.

Urbanisation as a process

When Berry wrote of 'cities as systems within systems of cities', he pithily expressed a connection between the organisation of each city and its place within a network of cities, structured both spatially and hierarchically.[33] The study of the forms taken by this connection has been a small but productive seam in both geography and history.[34] Cities are far more than containers; they are nodes in a network.

This interconnectedness of things is difficult to study and the external relations of cities have received less attention than they should, being confined primarily to the disruptive impact of migration on social structure.[35] As Southall shows below, the notion of the city as a container rather than a node supports a style of community study which, in its concern with stayers rather than movers, downplays the importance of the experience of a system of cities and focusses instead solely on the encounter of individuals with just one. The possibility, even expectation, of movement was basic for many nineteenth-century citizens. Sociological models often set up a threshold between traditional and modern society which is crossed with the move to a city. That is not the way urbanisation was experienced by individuals. The coming and going between town and country and the circulation of workers around a network of cities mean that place of residence on census night gives us a picture that cuts into a set of flows and misleadingly tempts us to read back modern expectations of relative stability of residence into the very different society of the nineteenth century.[36] In spite of the well-recognised mobility of the people of the nineteenth century, there is still a tendency to see movement as disruptive of social norms and none more so than movement between country and town.[37] Towns were connected with each other and with the countryside. Movement was the norm and experience of and familiarity with town life was far from being confined to those resident in town on census night. The uprooted peasant arriving in town straight from the country may, in one sense, be paradigmatic of the sociological contrast between traditional and modern society, rural and urban ways of life and even a convenient stereotype for the comparison of successive static maps, but it is an ideal type which, with its rural and urban 'boxes', presents far too static a picture of the mobility within and between the two, which was perhaps even more fundamental to social relations in cities than the trauma of the rural–urban transition.[38] The study of process must be more than the comparative study of sequential static pictures.

When one speaks of Britain becoming an urban–industrial society, this means the rural areas too. The framework of expectations in rural areas was radically altered by the continuous growth of cities. Landowners responded to growing markets for food, rural proletarians to expanding job opportunities.[39] Blaikie's paper serves to remind us that ideas about the countryside were extremely important to urban élites. As contemporaries tried to come to terms with the implications of the transitions they were living through, they constructed a concept of traditional society against which they might indicate the threats of the modern. They located this traditional society in some putative rural idyll and treated the political challenges of industrial capitalism as being solely due to the uprooting of workers from safe rural places and their exposure to the chaotic, corrupting and gargantuan cities.[40] Social relations were explained as spatial ones. Yet rural realities kept breaking in on these ideological fancies. Blaikie directs our attention to the place of rural myths in constructing urban bourgeois hegemony and to the empirical and political difficulties the bourgeoisie faced. The middle-class sense that the ground was shifting under their feet came as much from rural as urban realities. They perceived an urbanising *society* rather than just a growing *city* and, as sociologists were to do ever after, they saw the temporal shift as having a contemporary spatial analogue: here and now, there and then. From an ideological point of view, urbanisation was seen as a new state of things, defined in terms of its rural opposite. Raymond Williams' *The country and the city* is a brilliant exposition of the way the middle classes poured their own ideals and expectations into their perception of the contrast between town and country.[41] Williams showed that images of the countryside owed as much to the middle-class analysis of the causes of political and social tension in the towns as to the rural realities themselves. These images in turn bolstered the dominant diagnosis of the urban question.

Class and the city

At its worst, the urban ecological approach treated class as a set of pigeon-holes, or a scale from high to low, a filter through which data was passed before it was mapped.[42] This socio-economic grading, then, formed one dimension of differentiation within the city alongside others, of which ethnicity was the most important: the one based on occupation or profession and the other on birthplace, as given in the manuscript returns of the census enumerators. There is a risk that far too static a conception of class will result. Following Marx, Thompson has always insisted that we see class as a set of relations.[43] Classes only have meaning in their mutual interaction. Such relations can hardly be inferred from an occupational label. At the very least, static classifications need to be supplemented by detailed studies of local work practices and local cultural and political relations; the context in which class relations take on their meaning. Indeed, political and cultural practices and alliances are symptomatic of class relations in much the same way as is the residential segregation of occupational groups. All these sorts of evidence require evaluation if they are to be of use in describing and explaining urban class relations. The point is made clearly by Ward in his study of Leeds.[44] Does spatial segregation inflame class passions or not? At what scale should one look for meaningful segregation? Is the absence of some small part of the social spectrum from one or other part of the city, large or small, more significant than the greater range between that is found intermixed in the rest of the city? Is the frequency distribution of social status within an area more important than the average social tone of the area? Answers to these questions depend upon assessing the meaning attached to residential segregation by the different classes within the city. The construction of communities of interest was a cultural and political process and this was the context in which segregation took on its meaning. Bramwell's study of the uses of public space addresses this set of relations between community and class as does Withers' paper on Gaelic communities in Scottish cities.

Withers argues that ethnic groups were located by others as simultaneously a class and an ethnic group, the latter serving almost as shorthand for the former. Class relations prevailed within the ethnic group as well as between it and other groups in the city. This complicated the basis on which a community of interest might be asserted. Ward's useful discussion of an ethnic division of labour suggests one set of reasons why ethnically identifiable groups were discriminated against in labour markets, but the interrelations of class and ethnicity also go further than that.[45] There were fine calculations to be made by certain (usually better-off) members of an ethnic group between, on the one hand, attempting to integrate with the dominant classes in the city and, on the other, holding themselves apart as at least dominant within their own group. Questions of ethnic identity were thus clearly over-determined by matters of class. The role of ethnic contact networks in securing employment or accommodation for their fellows raises the question whether cultural homogenisation and separate development are really appropriate polar opposites when talking about the ways newcomers coped, or failed to cope, with city life. Powerlessness and assimilation might be characteristics that the bulk of ethnic groups shared with the rest of the urban working class. The significance of such things as ethnic segregation and community identity is not, therefore, something that can be evaluated on the basis of spatial data, but rather from attention to issues of class and the relations between people in given institutional and other social settings.

The images that classes held of each other crystallised certain aspects of class relations. Those images, as with the ghetto and the slum, occasionally incorporated a sort of moral topography. All sorts of metaphors were used to think about class relations and the spatial structure of the city. It is striking how important topographical metaphors were; quite obviously high and low, but also ideas about flow and stagnation. The well-ordered city would, it seems, have shared much with the physical geography of an improved piece of farmland or a country estate. In place of Cobbett's wen sucking people in, there would have been a growing, efficient, well-drained, landscaped town with its hydraulic systems in perpetual Chadwickian motion and its populace in healthy Boothian centrifugal flow.[46] As Kearns shows, these topographic metaphors were linked with biological ones. Ideas of contagion and environment were basic to the ways people wrote about class as well as disease. When a medical view of society was offered, there was an implicit conflation of social and biological relations. Class infused social epidemiology. Ideas of biological separateness reinforced a more primitive sense of the alienation of classes from one another. It is true that urbanisation transformed the biological basis of society and that, in particular, the demography of class was an important, though rarely studied, aspect of class formation; people found themselves not just in this or that occupation group but also as having this or that set of life chances. Class permeated the reality of the biological basis of society but also promoted a biological view of social and political relations.

Class and community

Community is something experiential, a process, a shared (if contested) set of meanings that attach to various social practices.[47] The urban ecological approach has looked at the distribution of institutions or at evidence of so-called interaction across space. These studies need to be supplemented by research on the creation of shared meanings. We cannot endlessly postpone the day when the study of the significance of marriage will enervate the frequency distributions of marriage distances, when the explorations of classes in interaction will agitate the maps of dissimilarity indices, when the study of the use made of cultural institutions complements the descriptions of their distribution. These things are regularly conceded in principle, but in practice there is a tendency to equate culture with mappable institutions. It is not, for example, the existence of churches that matters, it is the way that they fitted into and helped mould a whole way of life that we should study. Thus, Withers considers the place of churches in forging ethnic identity. These sites of cultural struggle have been of interest to urban ecologists often as *de facto* evidence of the existence of an ethnic culture without attention being paid to the significance they held for their putative clients.[48] Bramwell looks at the life of the streets as another arena of struggle. In asserting and expressing their culture in their daily use of these public spaces, working-class people had to contend with shifting possibilities for self-identification within their communities, but they had also to resist the efforts of external, middle-class forces to impose alien standards on public conduct. In exploring the forms and vibrancy of working-class communities, issues such as the middle-class civilising mission and broader working-class political radicalism have been much studied.[49] There has been much less attention paid to the construction, from within working-class communities, of shared values out of the repetitive routines of daily life. Yet, in certain respects, the tenacity of working-class communities, their responses to their exclusion from the hegemonic equation of the public sphere with that of the middle class and their resistances to direct assaults on their living standards through changing labour relations were all rooted in a militantly local culture that, as Bramwell shows, often successfully rejected and subverted external pressures. Both Withers and Bramwell focus on the active construction of communities from within, in ways that are often omitted from those spatial studies of interaction whose attention to community formation has rested on the study of marriage distances and the distribution of churches.

Southall and Bramwell draw attention in their chapter to different aspects of working-class communities. Southall considers community interaction in a system of cities and Bramwell examines the rootedness of working-class communities in local daily routines. Both studies emphasise the active construction of shared meanings out of shared experiences. These experiences were invariably structured by inequalities of class. Class and community are inseparable. The perceptions of and reactions to class inequalities were invariably framed by cultural factors. Market relations are embedded in cultural practices. Economic change takes on significance in the context of shared or contentious assumptions about the way things should be. Thus the contestability of change is about conflicting visions of what is right and proper and is also coloured by more obvious calculations of economic advantage. Craft traditions often provided an important focus for struggles around the labour process, but there was more at stake than the defence of wages. Craft traditions projected their own notions of community and of appropriate authority systems within those communities. The resistance to dilution of labour at work, by the employment of women or the unskilled, often carried over into an exclusion of women from organised political activity (except where issues of food or housing were directly involved, as they frequently were). The assertion of a shifting division between the rough and the respectable, as behaviour appropriate to particular contexts or, more aggressively, as labels for different social strata, could sometimes exclude poor, unskilled and immigrant from artisan-based ideas of community.

Class, community and urbanisation

In general terms, then, the theoretical developments in human geography and the burgeoning interests of social historians have influenced these essays in two main ways. First, class, understood as a set of relations, has been placed at the heart of our inquiry. Secondly, the study of how society is structured by class relations leads us to explore a range of cultural phenomena that express the way individuals signified to people of similar standing and to others the meanings they attributed to economic, political or demographic processes. The ecological tradition as expressed in much urban historical geography has not treated questions of class or of meaning very well. There are other important gaps in this dominant tradition, notably questions of gender, which these essays do not address. Neither do they add up to or consider any general theories of urbanisation in relation to our central concerns of class and meaning. There are, certainly, some general ideas about urbanisation and about the insertion of particular localities into broader processes of the spatial restructuring of capitalism that one might perhaps expect a set of urban historical geographers to consider, but they have not been explicitly discussed in this set of essays.[50] It is hoped that these essays will help others to explore these general ideas. Furthermore, we have made no attempt to recover and restate any clear principles of geographical method from these essays. Comparative approaches or concepts of scale, system, topography and spatial structure have never been the exclusive concern of the geographer and much of the best social and economic history shows itself equally familiar with them.

1

Biology, class and the urban penalty

GERRY KEARNS

The distinctive feature of the public health movement of the nineteenth century was that the remediable causes of epidemic disease were taken to be environmental and that this informed a policy of sanitary engineering. In contrast, in earlier periods the major epidemics were seen as principally contagious and the dominant strategies consequently focussed on quarantine and the control of the movement of those suspected of harbouring epidemic disease.[1] The implementation of the environmental strategy obviously raised political issues that touched class interests, since it proposed quite extensive interferences with private property.[2] Yet the environmental approach did not completely displace other conceptions of disease in scientific or in popular works. Disease, filth and contagion retained moral and class-based connotations that ran alongside the strict 'sanitary idea' of Edwin Chadwick and Thomas Southwood Smith. Similarly, a purely ecological account of the urban public health problem merely repeats but does not evaluate the ideology of the 'sanitary idea'. It is important to remember that various social and economic groups were affected to differing degrees by the insanitary urban environment. In order to win broad support for their proposals, the public health reformers presented the urban public health problem as one common to all classes and they criticised as inadequate a minimalist conception of the State that allowed public intervention only in the case of paupers, urging that disease was a common threat against which only collective rather than individual provision of sanitary services would suffice. However, they recognised that in addition to the contrast between town and country there were local causes that placed the folk of one part of the city, usually the poor, at somewhat greater risk than other citizens. The sewer and the slum were part of a moral as well as a medical topography.[3] To the extent that poverty might result from sickness the demography of the dangerous cities, the urban penalty, was an important component of the problem of urban poverty from both a class and a gender perspective. Poverty and disease were linked demographically, just as from an ideological point of view they were morally intertwined.

Sickness was frequently implicated in downward social mobility in the nineteenth century. How frequently is difficult to tell and assertions about its importance were so bound up in the general theories of poverty that they must be treated with scepticism. Dogmatically confident that the New Poor Law had eliminated poverty in England, Neil Arnott comfortingly assured Edwin Chadwick's 1842 Sanitary Inquiry that the persistence of fever in London directed attention to environmental factors and not poverty as the immediate cause of high urban mortality. It had fallen to Arnott to refute the claim of the Scottish doctor, William Alison, that the unhealthiness of cities was the direct result of the poverty of many of their citizens. According to Alison, sickness, even death, was the unacceptable price the poor paid for city living and the sheer numbers involved clear evidence that the New Poor Law had ignored rather than eliminated poverty.[4]

There is a great deal of qualitative evidence supporting Alison's position. Henry Mayhew's interviews with the poor of London are peppered with accounts of people being driven from regular employ by sickness. One pauper reported that he 'was forced to go to the parish...for I got ill and dreadful weak, and they gave me work on the roads', while another recounted that:

For six months I wasn't able to do a thing, and I was part of the time, I don't recollect how long, in Saint Bartholomew's Hospital. I was weak and ill when I came out, and hardly fit for work...My wife made a little matter charing for families she'd lived in...She was taken ill at last, and then there was nothing but the parish for us.[5]

Mayhew certainly got the impression that the Workhouse Test had succeeded in discouraging people from applying for poor relief, for many of his interviewees 'hate the thought of going to the "big house"' and will only apply for relief when 'fairly "beaten out" through sickness or old age'. Yet, the régime did not eliminate poverty and was seen by the poor 'simply as a punishment for poverty and as a means of deterring the needy from applying for relief'. One pauper commented to Mayhew: 'It seems to me as if in a parish a man must be kept down when he is down, and then blamed for it. I may not understand all about it, but it looks queer.' The reciprocal links between sickness and downward social mobility are difficult to untangle. Sickness may have driven many artisans into less demanding casual occupations, whereas unemployment may have so reduced others that they were prone to serious illness: 'I was very weak, you may be sure, sir; and if I'd had the influenza or anything that way, I should have gone off like a shot, for I seemed to have no constitution left.'[6] 'Anything that way' picked out the weak, the very people most likely to be exposed to it, the people living in overcrowded lodging houses or single-room families, with limited access to clean water and surrounded by their own and their neighbours' ordure. Structural unemployment, casual labour and a filthy environment reinforced each other in the Great Cities of the Victorian age. In the heartlands of the Victorian economic miracle, a wide health divide existed between the rich and the poor. Three aspects of this urban health divide are considered in this essay: the relations between the social structure and these biological consequences; the links between these life-and-death issues and the nature of urban poverty; and finally, the way these issues were treated by some bourgeois commentators, the opinions they held about the facts.

The health divide

It is extraordinarily difficult to determine class-specific mortality rates for nineteenth-century cities from the statistics collected by the General Register Office. It is also far from easy to interpret those we can calculate. First, the information about deaths must relate to the same groups as that on the population at risk. Surprisingly, it is the latter that is often weakest. Secondly, the socio-economic groups need to be reconstructed from the occupational labels actually used by the General Register Office. Many of these labels are far too broad in terms of social status, including both masters and men, traders and artisans. Thirdly, certain occupations had a distinct life-cycle, making direct comparisons between them unreliable. The professional occupations often retained their members into old age, while specific manual trades discarded their old and weak into catch-all callings such as 'general labour'. Fourthly, the various chronologies of the expansion of particular trades, such as commercial clerks, gave them a different age structure to occupations with more stable patterns of recruitment. Fifthly, the classification of occupations was distressingly unstable over time, with new specialisms, such as those employed on the railways, being hived-off, subdivided and variously regrouped at successive censuses; with traders treated as a separate category or put with the relevant sector of production on no really consistent basis; with the division between masters and men being occasionally reported and occasionally not. Sixthly, it is almost impossible to say anything at all about class-specific female mortality and certainly impossible to offer any reliable data on the matter.

Taking these difficulties into consideration, I have extracted from the Registrar General's Decennial Supplements for the second half of the nineteenth century information on a range of occupations that can with reasonable assurance be assigned to some of the socio-economic categories that historians, following Armstrong, are familiar with.[7] About one-quarter of males aged fifteen and sixty-five in 1901 can be allocated to the groups professional, white-collar, skilled or unskilled on the basis of occupational labels that are relatively consistent between census and registration material over the second half of the nineteenth century. As representative of the professional classes, barristers, clergymen and medical men have been included. The lower middle classes are represented by a relatively lowly group of white-collar workers: commercial clerks, insurance clerks, law clerks and railway officials and clerks. This excludes of course the shopkeepers, politically and economically among the most important members of the lower middle class. Following the conclusions of Crossick and the General Report on the Census of 1901, the following have been included in the skilled working class: cabinet makers, carpenters and joiners, coach makers, engine makers, gunsmiths, plumbers, painters and decorators, those engaged in shipbuilding and wheelwrights.[8] None of the occupational labels available seemed to adequately define a group of predominantly semi-skilled workers (Armstrong's largest group) and the following have been taken as broadly indicative of the unskilled: coal heaver, dock labourer, general labourer, railway labourer.[9]

Figure 1.1 appears to answer some of our needs. It shows for groupings of registration divisions the age-specific mortalities for our social groups in 1871.[10] I have shown the data for the age groups 25–34 and 35–44 as these will cover ages at which most groups had qualified and before many healthy artisans had been broken down to the status of general labourer. In one sense these data remain difficult to interpret because even the rural registration divisions contained some fairly significant towns and some of our groups, notably white-collar workers, were probably only found in towns; rather than comparing town and country, therefore, we are really comparing one set of towns with another. Nevertheless, the deterioration in the position of the unskilled as we move from rural to urban areas is probably significant, as is the wider health divide in urban in comparison to rural areas. It is also possible to calculate mortality rates for seventeen occupational orders in 1871, covering males aged twenty years and upwards, for London, for a group of seventy-nine other urban registration districts and for the 547 remaining districts. These orders conflate masters and men, traders and artisans, and are defined in terms of the materials used by each sector of the economy. In terms of social status they are virtually useless, but three bear some relationship to our social groups. First, 'unspecified labour', which was made up largely of general labourers, showed the following rates of mortality per thousand living for males aged twenty years and over in London, the other urban areas and the rest of England and Wales: 27.8, 30.5, 23.0. As above, we do not find the same range for the 'learned professions' (20.9, 22.0, 21.0) or those engaged in commerce, those who 'buy and sell', dominated by commercial clerks (respectively, 19.1, 20.5, 19.9). From these, admittedly imperfect, statistics I would suggest that it was particularly the poor and unskilled who bore the brunt of the urban penalty. It was the poor and unskilled (see Figure 1.2) who, 'fairly beaten out' and packed together,

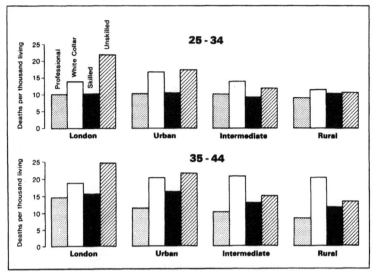

1.1   Age-specific mortality for social groups (males only) in the urban and rural
      registration divisions of England and Wales, 1871

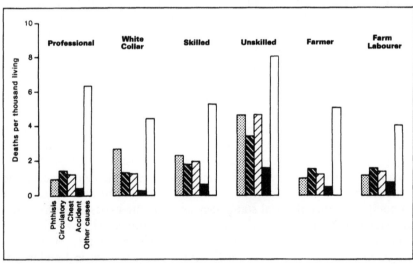

1.2   Causes of death for occupied males aged 25–64 in the different social groups,
      1900–1902

succumbed to phthisis (respiratory tuberculosis) and other diseases of the
chest or to diseases of the circulatory system.[11] It was they who led the most
dangerous lives. We can only document this dimension of inequality for the
end of the century, after the worst periods of urban epidemics, and we can

17

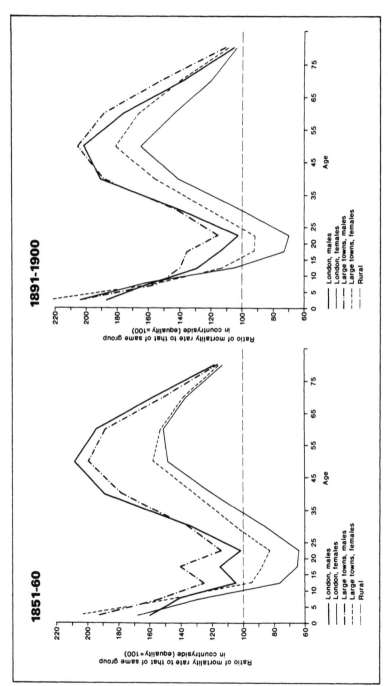

1.3 The relative mortality of urban and rural areas for males and females, England and Wales, 1851–1860 and 1891–1900

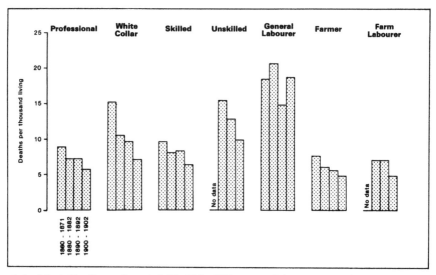

1.4   Mortality for males aged 25–44 in different social groups at selected dates, 1860–1902

only speculate that they might have reinforced the pattern shown here by phthisis. For the cholera epidemic of 1848–9 the rates of mortality (deaths per thousand living) among males aged fifteen years and over in the three social classes of gentry, tradesmen and mechanics in London were 4.8, 8.4 and 8.8 respectively.[12] Contemporaries certainly believed that the other urban fevers, such as typhus, had a markedly more class-specific character than cholera. Nevertheless, it was clearly in the towns and cities that the health divide was widest and the differences between the social classes most likely to take on a life-or-death character.[13]

All age groups, except infants, showed a marked improvement in mortality in the years 1851–60 and 1891–1900. Yet, as Figure 1.3 shows, the relative standing of the towns did not improve, despite the fact that age-specific mortalities in towns improved in line with national rates.[14] Thus, even after the great reduction in urban epidemics and in other contagious diseases, the urban–rural gap stubbornly remained. It is impossible to replicate this analysis for each occupational group, but the materials in Figure 1.4 are of some interest. Recall that professionals and white-collar workers cover predominantly urban occupational groups.[15] In 1860–71 professional and white-collar workers aged 25–44 had rates of mortality 49 per cent and 88 per cent of that of general labourers of the same age living in London. In 1900–2 the professionals had improved further and now had a rate of mortality that was only 30 per cent of that of the London general labourers. Far more impressive was the improvement in the mortality of the white-collar workers, where mortality was now only 38 per cent of that of the London general labourers. Perhaps one might conclude that the persistently high mortality of the London labourers reflects the balance between general environmental improvement with better drainage and sewerage for many and deteriorating housing standards for some caused by the chaotic overcrowding of the late-nineteenth-century metropolis. The dramatic improvement in the mortality of the white-collar workers might reflect the reinforcement of general environmental improvement by the growth of the suburbs and the new patterns of residential segregation that they allowed. It might even, quite recklessly, be suggested that the similar levels of cholera mortality among the tradesman and mechanics of London in 1848–9 reflected common environments and low levels of segregation at that date.

The urban penalty and the urban poor

Whereas 2.2 per cent of the male population over sixteen living in London were in receipt of poor relief on the occasion of the first truly comprehensive census of the poor (1906), the relief was received by 9.9 per cent of the unskilled compared to just 0.7 per cent of professionals and 0.9 per cent of commercial or business clerks.[16] At this date these proportions showed little difference between rural and urban areas. The fact that the same groups were reliant on poor relief as were characterised by high mortality is the completely unsurprising consequence of the place of sickness in downward social mobility. Not surprisingly, too, the elderly made up a very large part of this dependent population. Although many of the aggravating problems were much worse in urban than rural areas, the urban areas do not seem to have relieved a correspondingly greater share of their inhabitants. Taking as urban those places with a population of 50,000 or more in 1851, 0.70 per cent of the population in urban areas was in a workhouse on census night compared to 0.71 per cent of the rural.[17] In 1901 the respective proportions were 0.74 per cent and 0.55 per cent. There was a clear geography to this, with lower proportions in workhouses in the north of the country. At least in 1906, urban areas were more likely than rural areas to apply the Workhouse Test; about three-fifths of London's paupers over sixty were receiving indoor maintenance compared to one-sixth of those in rural areas. If, through rigorously applying the deterrent Workhouse Test, urban areas did not support the weak and dependent population, one might have expected them to. Nor did they offer specific medical relief on a particularly generous scale. In 1857 the ratio of indoor medical cases to workhouse poor was 1:3.0 in London and 1:2.8 in England and Wales as a whole.[18] For outdoor poor, the figures were 1:2.4 and 1:2.6. Consequently, the proportion of Poor Law expenditure spent on medical relief was much the same in London, 3.0 per cent, as in the country as a whole, 3.9 per cent, in 1857.

In addition to the relations between sickness, occupational mobility and a reliance on poor relief, there is a further important way that the urban demographic penalty shaped the problem of urban poverty and that is with the problem of widows. In European cities it was among men that urban death rates were highest. In rural England and Wales there was rough parity between the sexes and if anything it was women who, especially at mid-century, had the highest mortality. Between the age groups 20–25 and 45–55 (see Figure 1.3) the cumulative effects of urban environments and occupations led to a deterioration in mortality for both sexes relative to rural areas, at a slightly quicker rate for males than females. The heavy price paid by the children of the cities is also clear from these graphs, a situation that, in relative terms, was still worse at the end of the century than in its middle. Yet between fifteen and twenty-five, having survived the contagious diseases of childhood and before the beating out of a life of city toil, death rates showed a much smaller gap between urban and rural areas. At this point, city women are much better placed than their country cousins. Whether this reflects the safety of domestic service, the importance of industrial wages or the severity of rural female employment must await further research. There is also that surprising kink in the rates for males aged 15–20 in 1851–60 and still a suggestion of one for the males of large towns in 1891–1900. With the British data it is impossible to explore the most likely explanation for this, that single males (most heavily concentrated in this age range) formed the group with the highest mortality, thereby emphasising the importance of the family wage and the destructive nature of lodging-house living with its overcrowding and heavy drinking.[19] The grim reaper strode through the towns and cities of Victorian Britain with a particular eye for men. Throughout Europe the story was much the same; having cut down the men he left the cities chock-full of elderly widows. In London, at both mid- and late-century, three-fifths of the population aged sixty-five and over was female, of whom three-fifths were widows. Of the male elderly, but a third were widowers. In rural counties the disparity was rather less.[20] Given the importance of a family wage in the face of the vagaries of sickness and the labour market, it is not surprising that so many of the single elderly were in receipt of poor relief and that widows should have dominated a pauper population already heavily weighted towards the elderly. On Charles Booth's figures prepared for the 1909 Royal Commission on the Poor Laws, 14 per cent of the males of London aged sixty years and over were in receipt of relief (predominantly indoor) in 1906 compared to 17 per cent of females (divided evenly between indoor and outdoor relief). In the rural unions he distinguished (about 9 per cent of the population of England and Wales) 12 per cent of males and 18 per cent of females of this age were, in both cases, mainly in receipt of outdoor relief. Nevertheless these elderly males made up 19 per cent of the pauper population of London and 21 per cent in rural unions, while the females were 26 per cent of London's and 35 per cent of the rural paupers. The only way, albeit indirect, of documenting the importance of urban widows is to note that in 1906 4.3 per cent of widows aged sixty years and over were receiving indoor relief (an overwhelmingly urban system of relief for elderly women at this date) compared to 1.7 per cent of other women of the same age.[21] This suggests that about four-fifths of the elderly female indoor poor in England's cities were widows. Unsatisfactory as these statistics are, they indicate some of the ways that the demographic characteristics of the urban penalty structured the problem of urban poverty and shaped both what it meant to *be* poor and the conditions under which people might *become* pauperised.

Ideology and the biological view of class

The urban penalty, then, was highly selective as regards age, sex and class and these demographic realities partly shaped the problem of urban poverty as well as being a principal mechanism of downward social mobility. We are quite accustomed to speak of urbanisation as involving a fundamental change in the ecological context of society. Lampard has repeatedly emphasised the fact that the size and density of the population and the sort of infrastructure necessary to water and cleanse these multitudes involves a major shift in the relations between society and nature.[22] Such changes, though, had the differential consequences outlined above. In one sense, then, class differences took on a life-and-death character. Chevalier, at least, is convinced that, when considered alongside differences in patterns of migration, fertility and criminality, there are grounds for suggesting that class took on a biological character during the rapid growth of Paris in the first half of the nineteenth century. In the manner of their breeding, living and dying, large swathes of the working class, new to city life, rootless and desperate, were practically a race apart and they turned to crime or revolution to express politically and sociologically an alienation that was rooted in fundamental biological realities.

Chevalier laid great emphasis on the coincidence of political and demographic crises in Paris:

An instance of great significance is the appearance of cholera on two occasions, each coinciding with the severest crises – in 1832 and 1849. In both years, and even more in the dark days as the Restoration moved on to the July Monarchy, the biological drama cannot be separated from the economic and political drama; for the former reinvested the latter with its true causes and characteristics, which, in this instance too, were reduced to matters of life or death.[23]

The middle classes, unable to face the full truth about the nature of class conflict, insisted on a clear distinction between the anti-social dangerous classes and the more respectable labouring classes. Yet in their accounts of the dangerous classes writers such as Sue, Balzac and Hugo in fact gave accurate descriptions of the biological condition of the bulk of the working class, descriptions that anticipated the surveys of working-class districts in the writings of French hygienists. There are a number of problems with Chevalier's analysis. In relating revolution to the degradation and desperation of the new urban poor he perhaps paid too little attention to the explicitly political processes of revolution that brought artisans to the barricades in defence of established Parisian craft traditions. The advanced guard of the class war was not simply a revolt of the belly. Furthermore, there was more to the demographic crises than just the rate of city growth; after all London consistently outpaced Paris, was much larger and yet had both less cholera and less revolution.[24] However, Chevalier was surely right to insist that the biological consequences of city life during these periods of rapid urbanisation formed a central theme in the novels of the time and that there was a close relation between these novels and the propaganda of the public health movement. He was also right to insist that it is not only the facts themselves that matter but the opinions people had about those facts. He showed very powerfully that novelistic techniques were equally present in the writings of the hygienists as in the confessedly creative writing of the period.

I want to take up this suggestion of Chevalier and explore the circumstances in which Mayhew and Chadwick as propagandists, on one hand, and Charles Dickens as a novelist, on the other, treated class in recognisably biological terms. I want to show the similar ways they moved beyond merely reporting the demographic realities described above. For there is superficially an interesting correlation between the treatment of class in biological, even racial, terms and the dramatic moral tension of the 1840s and of the end of the century. In the former a public health perspective offered an environmental account of the removable causes of disruptive working-class agitation; in the latter eugenic notions attributed the same anti-social behaviour to the deterioration of the national racial stock through Irish influence and the inbreeding of the urban poor. Coleman has argued that the 'social novel' of the nineteenth-century city was fundamentally a creature of periods of acute political tension.[25] Concern with the state of the cities and of the urban poor, then, perhaps responded to political as much as demographic pressures. Consequently, although the theoretical arguments used were ecological or environmental, concern about public health also took on a political and class character. I want, by focussing on the treatment of questions of biology, demography and disease, to bring out the way classes were considered in the public health debate. The writings considered here are all by middle-class observers and I want further to suggest that there are class-specific psychological as well as political factors in the way poverty and disease were explored in them.

Victorian novels were saturated with descriptions of death and disease, often for pathetic effect. In Dickens, though, the contemplation of death was intended also to direct the readers' attention to God's judgement on those whose neglect of preventible causes allowed the death to occur. Alongside education, sanitary reform was one of the two principal crusades of Dickens' career. Thus when Dickens described the death of Jo, the poor crossing-sweeper in *Bleak house*, he presented a pathetic account of a weak innocent passing away while reciting the Lord's Prayer: 'Dead, men and women, born with Heavenly compassion in your hearts. And dying thus around us every day.'[26] Dickens wanted to renew his readers' sense of outrage at a state of affairs that long familiarity had inured them to. As Mr Morfin said in *Dombey and son*: 'I have good reason to believe that a jog-trot life, the same from day to day, would reconcile one to anything. One don't see anything, one don't hear anything, one don't know anything; that's the fact.' These deadened sensibilities Dickens contrasted with 'the ministry of Him, who, through the round of human life, and all its hope and griefs, from birth to death, from infancy to age, had sweet compassion for, and interest in, its every scene and stage, its every suffering and error'. Mr Morfin, indeed, tried to bear in mind the detachment and objectivity of the Final Judgement, a reckoning liberated from the self-serving claims of custom and practice: 'How will many things that are familiar, and quite matters of course to us now, look, when we come to see them from that new and distant point of view which we must all take up, one day or other?' From a propagandist's point of view, then, Dickens described the sanitary and attendant moral evils of the city in order to expand the 'contracted sympathies' of his readers.[27] A. S. Williams has given a fine account of the various ways Dickens did this.[28] Williams noted that Dickens contrasted, as did the public health pamphleteers, the development of society's mechanical productivity with the physical deterioration of its working classes. Dickens emphasised the fact that the contagion of disease re-established a bond between poor and rich that crass materialism wished to renounce. The alienation implicit in the indifference of the cash nexus left the poor rotting away in obscurity, fermenting a bloody political revolution. Finally, Williams claimed that it was Dickens' fear of the revolutionary mob that partly animated his concern for the poor. Clearly, we have in Williams' presentation of Dickens' work a biological description of the alienation of the working class that is superficially similar to that given by Chevalier for French novelists. This interpenetration of biology and class and a predominantly class-based conception of the public health problem is present in both the social commentaries of Chadwick and Mayhew, on one hand, and in the imaginative social novels of Dickens, on the other.

The physical deterioration of the working class was certainly a central concern of the broader public health literature. Both Edwin Chadwick and William Farr first approached the question of sanitary reform from a consideration of life insurance, asking what sort of liability society was implicitly accepting in allowing preventible disease and premature death and asking what loss of productive capacity or biological assets the poor health of towns caused. In his 1842 Report on the Sanitary Condition of the Labouring Population of Great Britain, Chadwick tried to illustrate this human waste. Noting that actuarial life tables were drawn up for particular places, he observed that 'very dangerous errors arise from statistical return and insurance tables of the mean chances of life made up from gross returns of the mortality prevalent among large classes, who differ widely in their circumstances'. Ignoring the effect of different age structures on his results, he went on to illustrate his claim with data on the average age at death among the various classes of different areas. In Manchester in 1837 the average age at death of 'professional persons and gentry and their families' was thirty-eight, while that of 'mechanics, labourers, and their families' was seventeen. In Rutlandshire the respective figures were fifty-two and thirty-eight. Both the poor and the rich were affected by the unhealthy towns. He then tried to assess the costs of this excess urban mortality in the numbers of widows and orphans thrown on the poor rates as a consequence. He asserted that 'of the 43,000 cases of widowhood, and 112,000 cases of destitute orphanage relieved from the poor's rates in England and Wales alone, it appears that the greatest proportion of deaths of the head of families occurred from...removable causes'. As a further illustration of this human waste, and with the true propagandist's eye for the striking comparison, Chadwick informed his readers that 'the annual slaughter in England and Wales from preventible causes of typhus which attacks persons in the vigour of life, appears to be double the amount of what was suffered by the Allied Armies in the battle of Waterloo'.[29]

Dickens emphasised the biological basis of the links between classes, but other commentators were just as keen to explore the biological basis of class differentiation. For example, the notion that social characteristics were carried in the blood gave rise to the idea that the differences between the classes could be passed, biologically, from one generation to the next, almost literally making the poor a race apart. The effects of the environment might be retained by successive generations of the poor. This is particularly significant because the environment was thought to have moral as well as physical effects. In the summary section of his report, Chadwick concluded:

That the younger population, bred up under noxious physical agencies, is inferior in physical organisation and general health to a population preserved from the presence of such agencies.
That the population so exposed is less susceptible of moral influence, and the effects of education are more transient than with a healthy population.
That these adverse circumstances tend to produce an adult population short-lived, improvident, reckless, and intemperate, and with habitual avidity for sensual gratifications.
That these habits lead to the abandonment of all the conveniences and decencies of life, and especially lead to the over-crowding of their homes, which is destructive to the morality as well as the health of large classes of both sexes.

That defective town cleansing fosters habits of the most abject degradation and tends to the demoralisation of large numbers of human beings, who subsist by means of what they find amidst the noxious filth accumulated in neglected streets and bye-places.[30]

There was a range of opinions on just how this culture of filth was propagated. Chadwick rejected Malthusianism and was silent on the anthropology found in writers such as Henry Mayhew.

Mayhew frequently made comparisons between the London poor and primitive tribes. For example, he depicted the civilised sedentary race of London as surrounded by a wandering tribe of street-folk. These less-civilised nomad peoples were characterised by their 'passion for stupefying liquors', by the 'looseness of [their] notions as to property', by 'the absence of chastity among [their] women' and by 'a greater development of the animal than of the intellectual or moral nature of man'.[31] When Mayhew wrote of plants being fed on sewage and thus producing valuable food out of disgusting muck, he commented that:

With that same wondrous economy that marks all creation, it has been ordained that what is unfitted for the support of the superior organisms is of all the substances the best adapted to give strength and vigour to the inferior. That which we excrete as pollution to our system, they secrete as nourishment to theirs. Plants are not only Nature's scavengers but also Nature's purifiers. They remove the filth from the earth, as well as disinfect the atmosphere, and fit it to be breathed by a higher order of beings.[32]

When he described London's scavengers, he wrote of their 'simple cooperation', or 'gang system' of working which, yet 'rude as it appears, is far from being barbaric', although 'the principle of complex cooperation in the scavaging trade exists only in its rudest form, for the characteristics distinguishing the labour of the working scavagers are far from being of that complicated nature common to many other callings'. Elsewhere, in writing of the £1½ million trade in second-hand articles in London, Mayhew follows them from aristocratic household through entrepreneurial Jewish used-clothes salespeople to:

[T]he very lowest grades of the street-folk – the *finders*; men who will quarrel, and have been seen to quarrel, with a hungry cur for a street-found bone; not to pick or gnaw... but to *sell* for manure.

There is a clear sense in Mayhew's work of a biological order to society with lower civilisations serving higher by feeding on and removing their expunged waste. For Mayhew, filth was at once physical and moral. For example, he reported that sewers were frequently used as places to hide such incriminating materials as the clothes from a murdered person: 'So close is the connection between physical filthiness in public matters and moral wickedness.' In one part of South London where a new sewer required to be built, a foul open sewer was flanked on one side by bone-boiling and other noxious trades and

on the other by houses with prostitutes calling out from them to passers-by:
'Perhaps in no other part of the metropolis is there a more marked
manifestation of moral obsceneness on the one hand and physical
obsceneness on the other.'[33]

I have already said that there is a superficial similarity between the
treatment of the causes of the revolutionary threat in Chevalier's account of
the French novelists and in Williams' account of the same matter in the
novels of Dickens. The fundamental difference is that, according to Chevalier,
the French correctly saw biological processes as the immediate cause of
revolution, whereas, for Williams, Dickens mainly used the metaphor of
revolution being a disease growing out of the degradation of the poor:

Since disease was most decidedly a fact of life, its employment as a metaphor would
have lent a sense of reality to the danger of revolution with the fear and suffering that
was associated with disease. Further, it would have served to remind readers that a
phenomenon once unknown in England [revolution] could yet be made known – as
in the case of cholera.[34]

There is certainly little evidence that public health reform was prominent
among working-class demands in either the 1830s or 1840s. Instead working-
class activity wanted a thoroughgoing reform and renewal of political
institutions, from which they ultimately anticipated all sorts of benefits,
including sanitary reform. In responding to Chartism with the dim perception
of a 'Condition of England' question framed in terms of the unsatisfactory
health of towns, the middle class was articulating its own concerns and
insecurity rather than listening to the demands of the working class.

There are a number of types of network in Dickens' novels. Characters are
related in terms of family, property/economy and morality/retribution.
Disease and death are common among the metaphors used in speaking of
these networks. *Dombey and son* is permeated by the metaphor of river Time
feeding the ocean, Death, and young Paul wastes away curious to know what
'the sea ... keeps on saying' until on his deathbed he realises what it is calling
him to. Death is presented in the sound of the waves, in the power of a train,
as old-fashioned, as a triumphant, indomitable monster and, at one point, its
name is London, as Harriet Carker reflects on the traffic on the highway:

Day after day, such travellers crept past, but always, as she thought, in one direction
– always towards the town. Swallowed up in one phase or other of its immensity,
towards which they seemed impelled by a desperate fascination, they never returned.
Food for the hospitals, the churchyards, the prisons, the river, fever, madness, vice
and death – they passed on to the monster, roaring in the distance and were lost.

Beneath this overarching metaphorical association of death, destiny and
London, there are other more restricted uses corresponding to the three sets
of relations already described. Speaking of blood relations, Dickens satirised,
in *Martin Chuzzlewit*, the claims of the Chuzzlewits to a pure and noble

lineage and, writing of Mrs Toodle's wet-nursing of young Dombey, he reassured his readers that 'Little Paul, suffering no contamination from the blood of the Toodles, grew stouter and stronger every day'. Yet, in the same novel, Edith Granger would not allow her grasping mother to take charge of precious, young Florence Dombey for, as her mother, Mrs Skewton, realised, Edith was afraid of the 'corruption and contagion in me', the 'grain of evil that is in my breast'.[35] In *Bleak house*, the slum property Tom-All-Alone's is personified as a sick person and we are told that 'there is not a drop of Tom's corrupted blood but propagates infection and contagion somewhere'. In a directly analogous manner, Richard Carstone's fate was sealed when he got involved with a contested will under consideration in the corrupt Court of Chancery, as Mr Jarndyce notes in speaking of the baleful effects upon Richard's character: 'It is in the subtle poison of such abuses to breed such diseases. His blood is infected and objects lose their natural aspects in his sight. It is not *his* fault.'[36] The rich try in vain to defy the Angel of Death with its retributive sword of miasma and contagion. Mrs Skewton, 'like many genteel persons who have existed at various times, set her face against Death altogether, and objected to the mention of any such low and levelling upstart'; and aristocratic Lord Dedlock disdained infectious disease as 'base contagion from the tainted blood of the sick vulgar'.[37] Yet Dickens presents their snobbery as futile in the face of a contagion that may visit upon the neglectful rich the price of their failure to govern in the manner which they claimed as their birthright:

As on the ruined human wretch, vermin parasites appear, so, these ruined shelters have bred a crowd of foul existence that crawls in and out of gaps in walls and boards; and coils itself to sleep, in maggot numbers, where the rain drips in; and comes and goes, fetching and carrying fever, and sowing more evil in every footprint than Lord Coodle and Sir Thomas Doodle, and the Duke of Foodle, and all the fine gentlemen in office, down to Zoodle, shall set right in five hundred years – though born expressly to do it.

The same contagion of pestilence, plague and spreading miasma stood as metaphor for the whole corrupt system of the Court of Chancery: 'Never can there come fog too thick, never can there come mud and mire too deep, to assort with the groping and floundering condition which this High Court of Chancery, most pestilent of hoary sinners, holds this day, in the sight of heaven and earth.'[38] This is the source from which sprang the sickness of the contested property and thus Richard Carstone.

In Dickens, corruption is both a physical thing (disease) and a moral category and it surfaces in both forms in relations of family and property. Likewise, filth is both physical and moral. The metaphorical weight is retained by the ambiguity of these terms. The frequency of these terms compared to other possible metaphors of social relations, such as mechanical efficiency or organic symbiosis, both as forms of expression and in the architecture of a novel about the corruption of tainted property, *Bleak house*, and one about the corruption of selfishness, *Dombey and son*, suggests at least a preoccupation if not an outright obsession with disease and death. Indeed in the novel in which London comes closest to being the central character, *Little Dorrit*, the secret at the heart of the metropolitan miasma is clearly disease and death. Schwarzbach summarises Dickens' view of the city in this novel: 'Its leitmotifs are an overriding atmosphere of suffocation and oppression, the surrounding wilderness of buildings and the dark and dank, mysterious river running through it. It is a view of London as a tomb'. The centrality of disease and death in Dickens' novels may be partly set to the account of his choice of the city as his subject. Yet the metaphorical ramifications of this focus cannot be explained in this way. Among these broader connections in *Little Dorrit* is the treatment of money as corrupting: 'Money in *Little Dorrit* seems so deeply involved with moral taint that even to have it is a sign of guilt...[I]t is often associated with death, the earth and even excrement...'[39] In pursuing these associations, Dickens was led to a moral topography and ecological anthropology very similar to that found in Mayhew. In *Our mutual friend* he wrote of the very scavaging business that Mayhew was surveying at about the same time. Dickens wrote of going 'down by where accumulated scum of humanity seemed to be washed from higher grounds like so much moral sewage, and to be pausing until its own weight forced it over the bank and sunk it in the river'.[40] Elsewhere, writing of the inhabitants of the riverside, Dickens echoes Mayhew's anthropology, describing 'a mud-desert, chiefly inhabited by a tribe from whom employment has departed, or to whom it comes but fitfully and rarely'.[41] The equation of money with excrement reinforces the moral lesson about the corruption of greed, but it also offers some prospect of salvation for the filthy and marginal groups in society, for, just as Mayhew speaks of the 'wondrous economy' that makes filth a source of sustenance, so in *Our mutual friend* the dust-heaps are a source of wealth, a way of wringing a living out of adversity and marginality. Here we probably come to the psychological undertones of the obsession with filth and disease in bourgeois views of the city and its poor.

The source of Mayhew's emphasis on nature's 'wondrous economy' may well have been Chadwick's insistence that all sewage be placed back on the land to provide further food:

For, as it is known that there is no waste in nature, so has it also now become manifest that, but for man's ignorant wastefulness, there need be no occasion to apprehend the pressure of population upon food; the means of its reproduction being ever proportioned to the amount of its consumption, and the limit set to the numbers of mankind being regulated not by pestilence and war, but by the power and wisdom, and goodness of Him who 'filleth all things living with plenteousness'.[42]

In a striking phrase Chadwick wrote in a private letter of being able in this way to 'realise the Egyptian type of eternity by bringing as it were the Serpent's tail into the Serpent's mouth'.[43] The Egyptian idea of eternity relates to the practice of placing the internal organs of dead Pharaohs in their mouths so that, by feeding on the system of their own digestion, they would always have nourishment in their long journey through infinity. For the Victorians, the swallowing of the tail also took on sexual connotations as an image of the conservation of bodily fluids through oral sex, offering the perpetual circulation rather than dispersal of sexual fluids and energies.[44] Here the common emphasis on bodily fluids and their correct regulation provides a potent link between sewage, sexuality and filth. It is no coincidence that the French hygienist, Parent-Duchâtelet, wrote principally on two topics: sewers and prostitutes. Indeed, he saw them as functionally similar in ensuring the circulation of bodily fluids, as Lécuyer reports: 'It is clear that a social physiology of excretion was the central theme in the work of Parent-Duchâtelet. The collection of fecal excretion by the sewers is indispensable to the functioning of the town, likewise the collection of seminal excretion by the prostitute.'[45] To be filthy, for a bourgeois, was to promiscuously discharge these life-supporting fluids. On this terrain the moral and physical concepts of filth meet and have their neurotic congress. Gay has written much the same about the crusade against masturbation, a potent force in any bourgeois sentimental education:

What made physicians, in company with their patients, so apprehensive about masturbation in the nineteenth century was that it seemed a pointless and prodigal waste of limited and valuable resources, leading, figuratively and often literally, to impotence. It constituted a loss of mastery over the world and oneself. The campaign to eradicate self-abuse was a response to that danger: a way of conserving strength and maintaining control, both highly cherished and maddeningly elusive goals in the nineteenth century.[46]

As Stallybrass and White demonstrated, the association of social marginality with filth in both English and French bourgeois writing came from a deep-rooted sense that to be bourgeois meant to set oneself apart from the scum, those without a clear sense of the importance of bodily regulation:

The bourgeois subject continuously defined and re-defined itself through the exclusion of what it marked out as 'low' – as dirty, repulsive, noisy, contaminating. Yet the very act of exclusion was constitutive of its identity. The Low was internalised under the sign of negation and disgust. But disgust always bears the imprint of desire. These low domains, apparently expelled as 'Other', return as the object of nostalgia, longing and fascination.

This exclusion, then, carried with it a strain and the denied bodily functions continually surfaced as an obsessive gaze at working-class supposed sexual promiscuity, at working-class abandon to the very bodily functions for which the proper bourgeois professed to have no name. In an arresting phrase they quoted from Babcock: 'what is socially peripheral is often symbolically central.'[47] Thus Mayhew's poor could be emblematic of the fundamental concerns of the bourgeoisie precisely because they were marginal. The slum and the sewer were central obsessions rearing up from the depths of their learned aversion to filth. When they cast around for an image of their political insecurity, then, it was perhaps no coincidence that biological models came so readily to the bourgeois mind.

Acknowledgements
I would like to thank the other contributors for their forthright and supportive criticisms. I must also thank Peter Compton, Pat Hudson, Marie Clarke Nelson, Mike Power, John Rogers, Naomi Williams and Bob Woods for their helpful comments. Thank you to Hannah Moore and Julie Holbrooke for assistance with data-processing, to Sandra Mather and Paul Smith for the illustrations and to Ian Qualtrough and Suzanne Yee for the photographic work. This study forms part of a study on the comparative urban public health of Britain and Sweden in the nineteenth century and is supported by the Wellcome Trust, for which I am grateful.

2

Public space and local communities: the example of Birmingham, 1840–1880

BILL BRAMWELL

Publicly owned space outside buildings represented a substantial proportion of the nineteenth-century urban landscape. Many activities took place in this space and these reflected and moulded the solidarities and divisions of society, including those of class, gender, age and neighbourhood. The task of reconstructing the connections between society and the use of public space is a considerable challenge.[1] While these connections are seen in most dramatic form in riots or in ceremonies and celebrations involving vast crowds, they are also evident in people's more regular and mundane routines, and there is a danger that historical research may focus in an unrepresentative way on the occasional, more colourful event. A further potential pitfall is to ignore the influence of the material conditions of capitalist relations and urbanisation on the social use of public space. This applies in some architectural discourse where the space between buildings is conceived in an idealist manner as the expression of the spirit of the age. It also applies in some work by historical geographers which simply establishes spatial correlations between residential social areas and descriptions of local social activities.[2] A central contention of this essay, however, is that it is important to consider public space and its social use in relation to how the subordination of labour to capital took place and was resisted.

A vital force shaping the most widespread public space of nineteenth-century towns – the thoroughfares – was the circulation of raw materials and partly and fully finished goods and of people travelling between their homes and places of work and consumer purchase. Of course this circulation was essential for capital accumulation and hence for the interests of capital.[3] The process of people circulating through public space in nineteenth-century towns also provided numerous opportunities for socialisation, with the resulting patterns of social activity both embodying and influencing social relations. Some of these social activities tended to be conducive to the interests of capital, but others were largely disruptive to its progress. The precise relationships of such activities to the interests of capital, however, are highly complex with often contradictory trends. For instance, the vitality and conviviality of much working-class street life could cement a sense of shared identity and common interests among the working class – thus enhancing the potential for class conflict – but could also provide the working class with mutual support and consolation which might diffuse such potential.[4] Class interests were often also behind attempts to change activities in public space in the nineteenth-century town. Middle-class preoccupations, for example, were a prominent force in attempts to modify aspects of working-class street recreations that were considered threatening to the values of respectability and industriousness. Yet such class intervention, intended to promote social stability, might itself heighten class hostility.

The complexity of and many contradictions in class-based capitalist society warn against a presumption that there was a linear progression to any sequence of alterations in the social use of public space. Unfortunately, some accounts of working-class culture over the nineteenth century do draw on a concept of an almost inevitable and unidirectional process of transition from a 'traditional' to a 'modern' society.[5]

The fundamentally class-based nature of the use of public space must also be understood in relation to the other social distinctions in nineteenth-century society, including those of age, gender and local community. All these dimensions of the patterns of social activities must be incorporated in a full explanation of the use of public space in the nineteenth-century town. Particular attention is paid in this discussion, however, to the reciprocal interdependencies between local communities and the social use of public space. Strong local community ties, for instance, might encourage social-isation among neighbours in public space and in turn reinforce neigh-bourhood attachments. Local communities, like class, are regularities in social relations that happened in history, are made and reshaped by people out of the pressures of the developing class society and have a localised geographical coherence (in this paper this is examined at a smaller spatial scale than the whole of a town). Although local communities involve many minor social transactions within a small area, they are also indissolubly part of the wider society and bound into its solidarities and divisions. Local communities are made up of conflicts and disunity as well as of what people had in common, and this was likely to be reflected in the patterns of socialisation in the public spaces within them.

It is useful to distinguish between those public spaces in nineteenth-century towns that served as major routeways, including the town-centre main streets and squares, and the less visible minor thoroughfares in residential neighbourhoods. Smaller public spaces, including courtyards and shorter streets, may also usefully be examined separately from such larger spaces as the public parks. This is helpful as the major routeways and larger public spaces normally lacked the intimacy of smaller neighbourhood spaces, with potential for different patterns of social use. The bulk of sustained analysis of activities in nineteenth-century urban public space has been of large assemblies of people – particularly the crowd and the riot – which usually occur in the major routeways, with the street life and society of more intimate neighbourhood space receiving far less attention.

The reconstruction of street life and community attachments in nineteenth-century towns requires attending to fragmentary and impressionistic sources such as newspapers. Evidence from newspapers requires handling with particular care, involving considered judgements as to whether inferences can be made on more general and commonplace activities and attitudes based on reports on events considered by contemporaries to be newsworthy. Major shifts in policing, for example, may have left most street life untouched, but newspapers were generally keen to emphasise changes or 'advances', thus providing a trap for the unwary researcher. Much of the press coverage of street life described disorderly behaviour or crime, but these reports can provide the clues on much more humdrum and widespread activities and values. Sustained immersion in such literary sources combined with continual critical interrogation can provide a basis to distinguish between the commonplace and the unusual and to discern regularities in people's values and behaviour. Although the task is difficult, even hazardous, it is essential to attempt it as the street life and local communities had an important bearing on the making of the working class; and such literary evidence is virtually all there is on certain aspects of these topics. It would be much more hazardous not to attempt the task as this may either sustain distorted interpretations based on insufficient evidence and research or lead to the neglect of these important subjects.

This paper examines the extent to which there was continuity or change in working-class use of two less visible public spaces – the courtyards and streets of primarily residential districts – and one more public space – the municipal parks. In particular the paper looks at this question in the context of the large provincial town of Birmingham during the period 1840–80. Over these forty years the population of this manufacturing and commercial centre grew from just over 180,000 to more than 400,000, maintaining it as the fourth largest town in England. Employment in the town was quite strongly concentrated in manufacturing industry. Some of the manufacturing trades employed large proportions of skilled workers and some used sub-contractors, who either worked in their employer's workplace or in their own small workshop. These workplace distinctions among Birmingham's population, and the question as to whether they were translated into marked social distinctions among the working class, have been much discussed by social historians.[6] The evidence for such social distinctions among Birmingham's working class is assessed subsequently in relation to the use of public space.

This discussion of working-class use of public space in Birmingham will be divided into three sections. The second and third will assess patterns of working-class activity in, respectively, the courtyards and streets and then in the municipal parks of Birmingham. The first section, however, reviews some of the existing research relating to the working-class use of public space in the nineteenth-century urban environment.

Society and public space in the nineteenth-century town

The few studies of the more intimate and secluded neighbourhood public space in English towns in the past tend to be concerned with a different social class or historical period to those of this paper or else only examine one particular form of activity in public space. For instance, F. M. L. Thompson and Donald Olsen have discussed the middle-class preference in the second half of the nineteenth century for a suburban environment providing a high degree of privacy, including such barriers from the outside world as gardens and hedges and distance from the worlds of work and the town centre. Leonore Davidoff and Catherine Hall have indicated that the emergence of these middle-class suburbs coincided with a separation in the spheres of activity of middle-class women and men, with women increasingly restricted to the private context of home, children and suburbs while men moved constantly between this private world and the public world of paid work, politics and the town centre.[7] The work of Jerry White does look at working-class activities, in this case in neighbourhood public space in London, but it explores the period between the end of the nineteenth century and 1939 rather than the mid-nineteenth century. He makes highly effective use of the oral testimony of former working-class Jewish residents of an East London tenement block and shows the importance of two public spaces around the building – the court and the balconies – for shared social interactions. Elsewhere he has reconstructed for the inter-war years the popular culture and relations between neighbours in Campbell Road, a poor and notorious street in North London.[8] While neither of these studies covers working-class experiences before the end of the nineteenth century, most other research has dealt with only a single type of social activity that took place in neighbourhood public space. Such work includes the examination of women's neighbourhood sharing networks in London before 1914 by Ellen Ross, the reconstruction of street-gang activity among urban working-class youths before 1939 by Stephen Humphries and the analysis of collective violence in urban working-class communities before 1914 by D. Woods.[9]

One of the very few studies that begins to give consistent theoretical emphasis to working-class use of neighbourhood public space in nineteenth-century towns is by Martin Daunton. In his research on the layout of working-class housing in Victorian towns, Daunton recognises there were connections between the internal and external spatial organisation of housing and the socially constructed threshold between public and private space. This leads him to look at neighbourhood public space in towns and, subsequently, to propose that the character of working-class use of this space was undergoing important changes over the nineteenth century. He contends that over this period: 'The space in the city between buildings...tended to become socially neutral, rather than social arenas in their own right.'[10] Similarly, he also states: 'The pattern of the late Victorian city was that people could assemble, but in a passive rather than participatory role, always under the control of a definite regulatory agency. The communal, *ad hoc* and participatory life of the early Victorian city had been severely curtailed.'[11] Daunton's undoubtedly valuable contribution is to highlight the potential significance of the street as space for socialisation in the nineteenth-century town. However, Daunton does not derive his chronology of decline in street life from original detailed research of primary sources. Perhaps this is inevitable as it forms only one theme of many in his general survey of the political economy of working-class housing.[12]

Certainly Daunton's proposal of declining vitality to working-class street life over the nineteenth century needs further careful empirical assessment. One reason for such scrutiny is that it is in stark contrast to the argument of another historian, Standish Meacham, that working-class local community attachments in towns only become strong at the end of the nineteenth century. He contends these strong attachments emerged only after mid-century because only then was there a slowing down of urbanisation, which led to reduced population mobility and greater residential persistence in working-class neighbourhoods.[13] The prominence of local community attachments for the urban working class late in the century is also indicated in Elizabeth Roberts' study of the oral testimony of residents of three north-west towns.[14] Of course such contrasting trends – with street life declining but local communities gaining coherence – may have coexisted, but it does indicate the need for further careful research. This paper sets out to test Daunton's proposals, in this case in the specific context of Birmingham over the mid-nineteenth century.

Public conduct in the more visible outdoor places of assembly and involving larger numbers of people has been the subject of rather more attention from historians. After all, crowds collected in the principal streets or on common land had clear potential to threaten and disrupt public order. Individual studies have taken as their particular focus such subjects as riots,[15] wakes,[16] fairs,[17] civic and royal celebrations,[18] the timing of crowd activity[19] and crime and policing.[20] Particular attention has been paid to middle-class apprehensions and to governmental interventions and personal crusades to tame public (and social) disorder. Most accounts agree that there were changes in public order over the nineteenth century, with a decline of violence and a calming of unruly disorder. There is less unanimity, however, over the sources of this change. Even among adherents to class-based interpretations of this trend, there is varying emphasis accorded to the relative importance of the imposition into the lives of the working class of an assertive bourgeois state, of modifications in class relations at work and of the continued autonomy and creativity of working-class culture. There is also an emerging debate about how thoroughgoing the changes in public conduct really were. This applies both to the degree to which boisterous, rowdy and violent behaviour characterised eighteenth-century public order and to the extent of change experienced over the nineteenth century.[21]

Of course the question of public order is pertinent to the less visible as well as the more prominent urban public spaces. Because of this, it is assessed in this analysis of mid-nineteenth-century Birmingham. But public order must not be allowed to dominate discussions of public space. Equal attention should be directed to the more typical and routine activities. This is particularly important given the case made in this paper that public violence and the most unruly disorderliness were not only less typical working-class uses of public space but also experienced an unusual degree of change over the period studied.

Analysis of the more visible public spaces in this study of Birmingham concentrates on the town's municipal parks. In common with other features of the urban built environment these spaces both reflected and contributed to the temper of class relations. Many among the middle class regarded public parks as positive inducements to the working class to withdraw from leisure pursuits that they considered eroded their sense of responsibility towards authority, the family and the demands of work. The present concern is less with the motives of influential people who sought municipal park provision than with working-class responses to these facilities. As H. L. Malchow hints for late-Victorian London, these amenities might be proposed in one spirit but used in quite another.[22] These discrepancies between intentions and outcomes have received scant sustained attention in published research, but they are a central concern of this essay. The discussion that now follows draws exclusively on a case study of Birmingham.

Courtyards as neighbourhood public space

In his discussion of neighbourhood public space Martin Daunton argues for fundamental changes resulting from the characteristic urban layout altering from courtyards to streets 'in the early and mid-Victorian years', although the precise timing is said to vary between towns.[23] He considers the courtyards of the early period more conducive to informal and intimate social use as they were so hidden from the general gaze and represented a shared dead-end space for relatively few houses. By contrast, the streets of the mid-Victorian years offered less potential for informality and intimacy as they were connecting routes open to public view and to regulation by outsiders. Together with this change in public space he describes the development of more privatised, domestic and family-centred lifestyles among the working class, indicating that these two trends were mutually reinforcing and discouraged informal and gregarious use of public space.[24]

In Birmingham, however, there was no such change from courtyards to streets as the characteristic neighbourhood public space between 1840 and 1880. New courtyard housing continued to be built in large numbers in Birmingham, until bye-laws introduced in 1876 required houses to have open space on two sides and front and back entrances, and there was relatively little demolition of old courtyards over the period.[25] While 35 per cent of households lived in court housing in 1851, this had risen to almost 40 per cent by 1871; and a survey of houses in 1875 revealed as many as 44.4 per cent lacked a back door, this being a characteristic of court housing.[26]

The continued intimacy and vibrancy of courtyard life over mid-century is not disputed by Daunton and is supported by the evidence for Birmingham. As late as 1890 it was said of Birmingham's courts: 'You cannot live in a court without knowing a good deal about your neighbours and their concerns, even without deserving the title of a gossip.'[27] Shared activities among courtyard neighbours feature in many contemporary descriptions. These are suggested in one account in 1880 of a 'maze of narrow alleys' that was said to 'swarm with people; children roll about the courts and alleys...while their mothers gossip over the washing of a few sorry rags, or sit on their dirty doorsteps'.[28] One incident of support among court neighbours is described in an account of the taking of the 1891 census, which explains how one elderly man in a court off Summer Lane 'was a sort of head man of the court and had very kindly filled in all the forms for the inhabitants'.[29]

It was particularly hard to avoid neighbours in a courtyard as the yard was the only access to the houses, was often the only prospect from the windows of the houses and contained the communal drinking water and toilet, washing and refuse facilities.[30] The use of these communal facilities could lead to disputes between neighbours: 'Mrs Brown and Mrs Perkins both lived in our court, and somehow or other had always been at open war. If Mrs Perkins wanted water from the pump, Mrs Brown was sure to be in particular want of it at the same moment. If Mrs Brown hung out her clothes to dry, Mrs Perkins' mop immediately wanted wringing in the yard, and the clothes were spattered with dirty water; in fact the court was kept in constant ferment.'[31] But sharing facilities could also encourage more friendly socialising. One account of the 'six weeks' wash' of linen items describes it

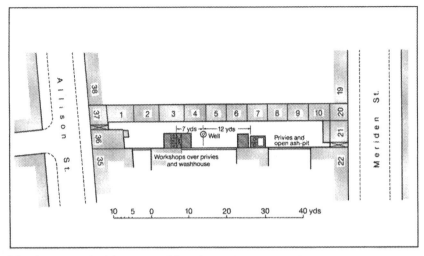

2.1 A courtyard with courtyard housing and communal facilities, off Allison
Street, Birmingham, c. 1871

*(Source: special meeting of the Council to consider the report of the Sewerage Inquiry
Committee, 'Birmingham Council Proceedings', 26 October 1871, appendix)*

as a ritualised social occasion among women using the washhouse: after
starting to dry the linen 'the presiding washerwoman having intimated that
it was time for the "regulars", the parties actively engaged adjourned from
the washhouse to the kitchen for the purpose of taking the customary
refreshment'.[32]

Intimacy among court neighbours was encouraged because the courtyards
so clearly constituted what Oscar Newman has called 'defensible space'.
Newman has argued that the subdivision of space outside homes by
arrangements in dead-end spaces and by arches or narrow entrances
encourages the creation of 'defensible space', with this territorial influence
being further promoted by windows that look out on the space to ease
territorial surveillance.[33] Certainly in Birmingham there was a clear
separation between front streets and the courtyards, with their 'close, arched
entrances, and the confined space within the courts themselves.'[34] The layout
of one courtyard, including the arrangement for the communal provision of
the only washing, toilet and refuse facilities for the courtyard residents, is
shown in Figure 2.1. Outsiders venturing through the narrow court entrances
were highly visible and likely to be carefully watched: it was described in
1880 how the 'gossiping women' who sat on their doorsteps in the courts of
Duddeston 'stare at such a rare phenomenon as a stranger invading their
region, hazarding guesses as to whether he is "a School Board", "a tally-
man", or "a 'tective"'.[35] The sense of territorial influence that existed in

Birmingham's courtyards emerged because of the interplay of the attitudes and practices of the residents with both their physical and social environments.

Territorial attachments also extended beyond the courtyards into the nearby streets where many activities had to take place. Courts had inadequate basic light and space, virtually no facilities such as shops and pubs and insufficient range of social contacts for them to be socially self-contained. For instance, the deficiency of light is cited in 1871 to explain why court residents lounged at street entrances to their courts: 'The courts are dark, cheerless and uninviting...One does not wonder at the poor people creeping out of their dark abodes to bask in the warm and pleasant sunshine.'[36]

Neighbourhood public space and local communities

There is relatively little support in the Birmingham context for Martin Daunton's proposal that the streets were 'socially neutral' or that people were using them 'in a passive rather than participatory' manner by the later 1860s or 1870s.[37] Indeed, what stands out for this town throughout mid-century is the strength and vitality of social activity in both the streets and courts in working-class neighbourhoods. This social interaction among neighbours was commonplace in working-class neighbourhoods, irrespective of the precise residential mixture of manual social strata. For instance, in Hockley, an inner suburb of fairly good working-class housing (with what were described as 'small faded gardens'), it was common in summer for the 'inhabitants [to] air themselves in their shirt sleeves, outside their special castles, and the matrons exchange recipes and small talk in a like open way'.[38] In the more central and more modest working-class Graham Street, it was described how 'on Sunday morning, if the weather be fine, some of the inhabitants came out on the door steps to air themselves'.[39] Similarly, in the streets of the far poorer central slum district around Aston Street the 'men, sucking lazily at pipes, lounge idly in the sunshine amusing themselves with the choice words in which a certain type of "the Birmingham man" is prone to indulge'.[40] Assessment of the extent of continuity and change in street life in working-class districts between 1840 and 1880 follows subsequently. First, however, the discussion explores some general characteristics of street society.

Frequent chance encounters in the streets of densely crowded working-class areas encouraged familiarity and conviviality among neighbours. Intimacy among neighbours meant they shared experiences and social networks, all of which reinforced people's sense of their own social relevance, including giving people a sense of importance, status and success in their own local community.[41] This had considerable significance as the local community was one of the few contexts in which working people could feel their social relevance; in the wider society they had little wealth, power or prestige. The desire for acceptance and status among neighbours has much historical importance as it is firmly bound into the commonality of working-class values and behaviour. Some of the interdependencies between neighbours occasionally emerge from the historical record when there were displays in the street of solidarity and support for a neighbour. Occasionally there was such a display of support during a visit by the bailiffs. When bailiffs attempted to take possession of a widow's furniture in one modest working-class street 'a large number of persons residing in the neighbourhood congregated round the door, and commenced to annoy and interfere with the bailiffs'.[42] After the bailiffs had succeeded in removing furniture from one house in a poorer street a 'mob' stopped the van in a near-by street, took out the furniture and broke it up.[43] Another account of reactions to the bailiffs described how 'sympathetic and indignant neighbours ... give us an effective parting groan. The women were the most troublesome. We had a mob of women and children round us in one street, when an old lady, boiling over with indignation, hung on to the bridle of our high-mettled steed.'[44]

An active street life encouraged the neighbourhood intimacy that offered individuals local social relevance, but it also had drawbacks. For instance, it meant that quarrels between neighbours sometimes took place in the street and so were highly public. These public confrontations often flared up without warning around such matters as the behaviour of children or damage to washing.[45] The street was also a place where very occasionally individuals who exceeded the bounds of acceptable behaviour in a neighbourhood might be subjected to demonstrations of collective public censure. One dramatic instance took place when a neighbour remarried only three months after his wife's death, which 'aroused the indignation of the residents of the neighbourhood' and a crowd of 'men, women and children ... pelted them with eggs and slush, threw old boots and rotten fruit at them, and assailed their ears with language gross and vile'.[46] Another similar street demonstration followed the marriage of a couple with a large disparity in their ages. The bride and groom were 'received with ironical cheers, and scores of lady-loungers favoured them with "pieces of my mind"'.[47]

Such dramatic public spectacles of neighbourhood disapproval took place only occasionally but served to highlight the risks of breaking neighbourhood behavioural standards.[48] Another encouragement to conformity with local codes of acceptable behaviour was fear of adverse local gossip, with the street also being prominent in the flow of rumour and opinion. Returning to the example of visits by the bailiffs, it was described how, after the departure of their van, 'old women ... stand at the street corners for an hour or more, discussing the scandalous event, every little child knows of it', and how the father of the family that was affected would 'be very coldly looked upon for the next week or two, as he slinks shamefacedly in and out of his house'.[49] The rapid spread of gossip is suggested in an account of events after a woman died after being pushed to the ground by her husband: 'A little time sufficed to make the occurrence known in the neighbourhood, and the excitement manifested by the inhabitants of the vicinity was great in the extreme ... The curiosity of the inhabitants had in the meantime progressed to a very high pitch, and a large and noisy crowd collected round the house of the deceased.'[50]

Unpopular neighbours could be subjected to public ridicule or abuse and sendings to Coventry in the street. For example, a woman had the front windows of her house smashed by youths living in her neighbourhood and she 'attributed the abominable treatment of which she had been the victim to the fact that she would not associate with her neighbours'.[51] In another instance, a man mocked the reformed behaviour of a neighbour – he had begun to attend adult Sunday School classes – by 'point[ing] him out to the neighbours' as he walked along the street.[52] Fear of all forms of neighbourhood sanctions in the streets could suppress individualism but also checked such abuses as wife-beating or excessive drinking. Concern not to lose status in the local community by being subjected to such street sanctions no doubt encouraged the commonality of much working-class behaviour. This neighbourhood intimacy had even more general social implications as it could encourage both the parochialism and consolations of local community concerns and the solidarities of working-class culture, which promoted collective demands against the interest of capital.

Yet there were variations in the extent and character of street involvements for individual working people. For a few, their less intimate dealings with neighbours resulted from poor social skills. But some felt that social distancing from the full gregariousness of the local community was necessary to maintain standards; they saw certain neighbours as a threat to their respectability. One working-class woman who was concerned about living in a poor district in Deritend was described by a mission worker as 'a respectable woman ... Not liking too much "neighbouring", she is accustomed to keep her household affairs in her own hands.'[53] Another mission worker told of families who 'by-and-by becoming too respectable to dwell any longer in the dark courts in the town ... migrate to the fashionable quarters of Sparkbrook, Small Heath or Aston. Because, as they say, "now we do not like our children to hear the bad language here; we never used to notice it before, but we cannot live amongst it any longer"'.[54] The great majority of working people, however, generally avoided being too stand-offish or snobby with their neighbours in the street, except perhaps on occasions of street violence or rowdyism, as to do so risked being completely excluded from their street and neighbourhood society. A glass manufacturer in the town noted that the élite of his skilled glass blowers and his less-skilled glass cutters 'may speak perhaps in the streets if they happen to know one another', but at work, 'though working on the same premises, one never passes into the workplace of the other. It would not do.'[55]

There were also distinctions in street life between a few highly impoverished neighbourhoods immediately behind the town-centre main thoroughfares and other neighbourhoods. The former were more prone to excitable social relations between neighbours and to rowdiness. One of these was Green's Village, described as 'crowded with human beings – ragged unwashed children, screaming, fighting, swearing women'.[56] These districts were some of the town's worst slums and housed large numbers of the unskilled in such casual work as street trading and labouring. Some contemporaries identified the periodic eruption of noisy rows and brawls in streets in some of these districts as specifically 'Irish rows'.[57] According to an account of one such incident in Park Street: 'last evening the classic neighbourhood of Park Street was in an unwonted state of excitement consequent upon a lively row in a yard principally inhabited by families of Milesian extraction [Irish]. Bricks flew about in all directions, women screamed and tore each other's hair, and an admiring crowd looked on, applauding with voice and action the exciting "mill".'[58] This unusual level of boisterous and sometimes violent street activity was influenced by the wider culture of many of the residents, but it was heightened by the presence of numerous poor lodging houses and by drunken visitors from the near-by town-centre pubs.[59]

Continuity and change in neighbourhood public space

Pressures for change in working-class courtyard and street life came from several directions, including regulation of the streets by the police and the opening of municipal parks as alternative public spaces with presumed clearer local government supervision. There was no necessity, however, that these pressures would cause profound changes in street life as in practice much depended on working-class responses to them. Modifications in the use of public space could also emanate from shifts in working-class culture, such as with the emergence of more privatised and family-centred values. Clearly the extent of changes in Birmingham's street life between 1840 and 1880, and the reasons behind this, demand further sustained study.

The power of Birmingham's police to regulate street activity represented one potential pressure for change.[60] This potential did not increase from a growing police presence, however, with the town's authorised full complement of police representing one law enforcer for every 718 residents in 1857 and one for every 716 residents in 1876.[61] In addition, attempts by the police to suppress street life were quite often unwelcome, resulting in evasion and, for some, in deliberate and even aggressive acts of defiance and hostility. Although the police sometimes intervened in such street activities as casual assembly and loitering, noisy private quarrels and boisterous drunken behaviour, these were widely considered to be non-criminal matters that did not warrant police interference.[62] Although many people sought to avoid getting involved in scenes of drunkenness and noisy quarrelling in the street as non-respectable and demeaning, such incidents were still viewed as matters of concern only for the individuals, their families and neighbours and not for the police. Many considered them a feature of life in working-class neighbourhoods and felt it was unrealistic to expect this to change.

If the police did intervene in street activities of this kind, some responded by carrying on – trusting to luck to avoid the police, carefully watching for their presence or simply ignoring their reprimands. The young Will Thorne showed obvious tenacity in continuing to run in street pedestrian races in the face of police harassment: 'Often the police would interfere with our sport and threaten to arrest us, but in spite of the danger of the law we continued our contests, so keen were we on the competition of it.'[63] One common ploy among youths playing pitch-and-toss was to use look-outs, as described in 1871: 'As usual with these Sabbath breakers, there were a couple of advanced sentries stationed round the camp, who gave the alarm on the appearance of the "bobby".'[64] But occasionally some reacted to police interference in street activity with violence directed against the police, including assaults and attempts to 'rescue' prisoners taken into custody in the street.[65] During 1867 there were as many as 1.16 police assaults for every policeman in the town, and 0.79 in 1877. This represented 0.9 police assaults per thousand population in 1861 and 1.08 per thousand population in 1877.[66] Such police assault statistics may be suggestive of trends in general police–public relations in the streets. But in Birmingham there are few consistent changes between 1859 and 1880 (when figures are available) in the ratio of assaults to either total police numbers or total population, and hence there is no indication of greater acceptance of, or resistance to, police interventions.[67] Although working people often disliked police interference in street life, it is also the case that most disapproved of violence against the police or at least avoided involving themselves in it. Barbara Weinberger's research on crime in nineteenth-century Birmingham compares the social characteristics of those charged with police assault with those of the general population and identifies an over-representation of unskilled and semi-skilled manual workers, young men and Irish, and an under-representation of skilled workers.[68]

This common dislike of police intervention in street life encouraged policing that was often cautious and sensitive to neighbourhood responses.[69] Of course this style of policing did not please everyone. One correspondent to the local press complained in 1863 that 'so little notice is taken of the fighting in the streets of Birmingham by the police'.[70] The *Town Crier*, a local satirical magazine, mockingly suggested a new bye-law for the town: 'No policeman shall, except under severe pressure from some ratepayer, be seen near or interfere in any sort of disturbance in the streets... or hurry beyond the regulation pace to such a scene.'[71] The threat of violence against the police was greatest in certain of the poorest slums, and the police were especially wary of entering these neighbourhoods. In 1873 one newspaper even argued: 'It is now a well-known fact that some parts of the town on Sundays are monopolised by the roughs, who amuse themselves with gambling, throwing brick-ends at houses... and rescuing prisoners from the custody of the police when an opportunity presents itself. Consequently those parts of the town are actually given up to mob law.'[72] More generally, however, policing the streets in all working-class neighbourhoods tended to be responsive to the wide acceptance of a traditionally vibrant street life.

Police intervention did contribute, however, to substantial modifications in a few neighbourhood street activities over the mid-nineteenth century. Douglas Reid has ably demonstrated that in Birmingham policing was an important contributory factor to the decline of wakes as popular neigh-bourhood street festivals, including the taming of the bawdy behaviour previously closely connected with them.[73] Police regulation also played its part in a diminution in cases of 'cruel' sports – such as dog fighting – and of prize fighting, both of which had sometimes taken place in more secluded public spaces in the town. By 1871 it was stated that dog fights were 'almost impossible (thanks to the vigilance of the police) to "bring off"... in town' and also that prize fights were very rare, even in remoter rural locations outside the town, so that 'the ring is almost a thing of the past'.[74] But the wakes, cruel sports and prize fights were unusually easy for the police to monitor and then suppress. This was because they involved deliberate planning in advance, they took place only occasionally and at a pre-arranged time and place and they often drew large numbers of people from a wide geographical area. By contrast, most other working-class street activities took place spontaneously, had an impromptu character and usually brought together only small numbers of friends, neighbours and passers-by. Even among street incidents involving violence, very few involved prior planning to help police detection; most street brawls, for instance, began spon-taneously in the heat of an argument. One typically impromptu fight involved two men who 'were well known to each other, and had previously lived on amicable terms [and]... were drinking together at the Rollers' Arms, when a quarrel arose between them. Words beginning to run very high, they decided to adjourn to a piece of waste ground a short distance from the house, in order to fight it out.' As was common with street brawls, this was treated as a matter of general concern among neighbours and passers-by and soon a 'large crowd' had assembled to watch.[75]

Changing attitudes among sections of the working class also produced some change, and this was especially pertinent to the decline of wakes, cruel sports and prize fights. Over the nineteenth century rising numbers of the working class avoided these street activities as they were identified with deliberate, planned excesses of disorderly, violent or cruel behaviour. For instance, it was argued in 1871 that altered sensibilities had influenced the downward course of dog fighting: 'for the credit of the dog fancier of the town be it said, they set their faces against such wanton cruelty, almost to a man'.[76] While many were coming to reject such planned excesses, a smaller but growing number also began to distance themselves from the spontaneous incidents of violent or very rowdy street behaviour. One possible repercussion of these changes among the working class, together with police activity, was that over the century street assemblies of neighbours to censure individuals for breaking local codes of conduct were becoming less likely to involve physical assault. It was suggested in 1851 that 'Birmingham is not without its code of mob law', its 'kind of Lynchism', but this was compared favourably with earlier periods 'when the mob ruled almost without check in the town, and punished offenders against its laws...by summary ejection from neighbourhoods accompanied by the horridest tintinnabulary uproars, known as "tin-kettling"'. It was also argued that: 'The chief exhibitions of this nature...have been left behind in the march of intellect; or what is more probable, suppressed by the new system of police, but the sentiments in which they originated still remain in the popular mind, and are occasionally productive of strange results.'[77]

This is not to suggest that the distancing of some people from violence or excessive rowdiness profoundly reduced the vitality of street life. It did not. After all, most street activities were not popularly associated with these characteristics and so they continued much as before. Even incidents of public censure of individuals by neighbours had always been more likely to involve ritual humiliation – which was more generally acceptable – than physical harm.[78] The spontaneous nature of most disorderly street incidents also meant it was very difficult in practice to avoid them totally. There also remained some people easily drawn into rowdyism, and many others who might join in when influenced by drink or the heat of the moment. As many working people drank, on occasion quite a lot, they could not wholly exclude the possibility of doing things that in other contexts they might find shocking. Similarly, there continued to be certain impoverished districts where street brawling was more frequent and more likely to involve whole families and attract large crowds.[79] Neither was the decline of wakes, cruel sports and prize fights to herald the triumph of a new passive street order. The emphasis in the historical record on the decline of these few street activities reflected their colourful nature as well as middle-class concern to demonstrate improvements in public order, but this must not be allowed to draw too much attention from the considerable continuities in street life. Above all, a communal and participatory street life was ensured by its continued importance to neighbouring and local community.

Continuity and change in juvenile street life

Daunton appears to accept there were many continuities in the vitality of street activities for young children and youths over the nineteenth century. He argues that 'the more active forms of the early Victorian city lacked respectability and were ever more controlled, relegated largely to a juvenile sub-culture'.[80] The significance of continuities in juvenile street life must not be underestimated or too marginalised, however, as juveniles were particularly numerous – in 1871 as many as 47 per cent of Birmingham's population were aged under twenty years – and because of the potentially formative influence of social experiences in early life.[81]

There is little evidence for Birmingham of change over the mid-century in the vigour of either innocent or more boisterous juvenile street recreations. The many continuities in children's innocent street play were encouraged by the crowded houses, which meant parents were often relieved to propel their offspring out of the house, and by the attractions of busy streets.[82] It was described as late as 1896 how: 'Girls of the working class ... are from their earliest years accustomed to the streets, where they play with their companions.'[83] The resilience of more exuberant and sometimes rowdy street activities for significant numbers of older youths is also evident.[84] Such street recreations, including running and walking races and boxing contests, were a regular routine for the young Will Thorne in the 1870s. For instance, he described how 'my companions and myself started boxing and training at the street corner very nearly every night ... We used to pummel each other, and arrange matches between ourselves, each contributing towards a stake, which, however, was rarely more than a shilling or two.'[85]

Street gambling continued to have a large following among working-class youths in the town throughout the mid-century.[86] As well as small-scale assemblies, with a single gambling cloth or board, there were also large gatherings – sometimes involving over fifty or even hundreds – on less conspicuous waste-ground or churchyards. One well-known meeting place for gambling was off Suffolk Street, where 'large parties of youths and boys, too, assemble daily, nay almost hourly, Sunday as well as week-days, upon the waste-ground now cleared for the railway, to play at pitch-and-toss'.[87] Over the period there is also sustained newspaper reporting of the involvement of numerous working-class youths in street gangs. Indeed, there is an increase in press reports of street-gang incidents during the 1870s, but it is unclear whether this reflected a real trend or only heightened press and middle-class concern for this form of disorderliness.[88] Most of the street gangs were composed of older adolescents, a group in a difficult social position, neither having the interests of young children nor being fully accepted into adult society.[89]

Participation in street gangs involved idling in groups in the streets and, on occasion, street fights with rival gangs from neighbouring streets. This provided an outlet for adolescent energy and enabled youths to gain a form of power and prestige in their own local community. According to one commentator some adolescents in the town's street gangs in the 1870s were known as 'peaky blinders' because of their distinctive style of clothes, describing their 'bell-bottomed trousers secured by a buckle belt, hob-nailed boots, a jacket of sorts, a gaudy scarf and a billy-cock hat with a long elongated brim. This hat was worn well over his eye, hence the name "peaky blinder". His hair was prison cropped all over his head, except for a quiff in front which was grown long and plastered down obliquely on his forehead. He usually belonged to a "slogging" gang, and terrible fights took place between rival gangs.'[90]

Fights between street gangs often involved territorial conflict between neighbouring gangs. This reflected their assertion of masculinity, their desire for excitement and their own particular sense of identity and pride in their street and local community.[91] One illustration of this territorial form of contest is provided in a report of a court case after a fifteen-year-old boy had been attacked by thirty youths: 'There are two sets of boys in the habit of fighting each other. Joyce belongs to the Park Street set, and I belong to the Milk Street set... Some of the boys on both sides carry knives and use them. Some carry buckles, and there is stone throwing... The lads in different streets are in the habit of what they call slogging one another.'[92] Although rowdy street-gang activities annoyed many local residents the local loyalties of the gangs could be widely drawn – one respectable resident commented of his neighbourhood street gang: 'Perhaps as we lived on his doorstep, we were treated as members of the gang by courtesy or adoption. I had to go home at all hours... and the local bullies would always give me curt nods of comradeship.'[93]

Although the characteristic forms of juvenile street life in Birmingham continued largely unaltered over mid-century, the expansion of schooling during the 1870s did mean that their timing was increasingly restricted within out-of-school hours. Prior to the 1870s many children had the potential time during weekdays to frequent the streets. An investigation in 1868 found as many as 43 per cent of the 37,122 children aged five to fifteen years in the survey were neither attending school nor in regular employment on weekdays (with 39 per cent at school and 17 per cent at work). Even among the ten- to fifteen-year-olds as many as 38 per cent had no regular job and did not attend school on weekdays (with 25 per cent at school and 37 per cent at work).[94] This situation was to change following a local bye-law of 1871 requiring school attendance for all five- to thirteen-year-olds below a prescribed educational standard.[95] To accommodate these children the local School Board opened the first board school in March 1873 and had built as many as twenty-eight with places for 28,787 children by 1880.[96] Children aged between five and thirteen came under the surveillance of School Board attendance officers from 1872, with seventeen such officers by 1876.[97] But, despite this, numerous five- to thirteen-year-olds were still free to play on the streets on weekdays in the late 1870s. One reason for this was the shortage of school accommodation in several districts of the town.[98] In 1878 the public elementary schools had room for only 52,171 of the 61,118 three- to thirteen-year-olds on their books.[99] There were also too few School Board attendance officers for adequate monitoring of school attendance, and a regular census of all children was kept up only in the poorest districts.[100] Furthermore, parents saw that only parents with children who were very persistent non-attenders were taken to court and that fines were often well below the maximum penalty.[101] It is perhaps unsurprising that the average attendance at the town's public elementary schools over one week in May 1878 was only 71.2 per cent of children on their books.[102]

Public space and municipal parks

Another of Daunton's proposals is that the opening of municipal parks in towns during the nineteenth century added to the control of public space by definite regulatory agencies. He suggests this contributed to the trend over the nineteenth century whereby people increasingly used public space in a passive rather than a participatory manner.[103] Certainly the municipal authorities providing parks often saw themselves as purveyors of well-ordered, passive and even reflective public behaviour. The hope was that parks would promote more disciplined behaviour among the working class both during their visit and, as a result of the social training they provided, in their subsequent use of other public spaces.[104] But it should be emphasised that these more visible public spaces might be proposed with one set of intentions but in practice be used quite differently by the working class. Similarly, while the parks could be intended to reform the complete lifestyle of their working-class visitors, including their approach to street life, their actual social influence might be far less significant. The intention here is to assess the degree of impact of municipal parks on the general use of public space in Birmingham.

The expansion in municipal provision of parks in Birmingham during the nineteenth century was a manifestation of the slow realisation among the middle class that they had some responsibility for, or self-interest in, the health and welfare of the working class. It was also a response to working-class demands for recognition and reform, with working people perceiving that there were certain advantages arising from such municipal amenities. In the public dialogue among the middle class in support of municipal provision of parks three sets of arguments figured prominently.[105]

First, it was suggested they would be beneficial to working-class health. For instance, one Birmingham newspaper argued that municipal parks were needed as the town lacked 'a rood of land left in which our toil-worn and debilitated workmen can get fresh air, or find breathing room for their contracted lungs'.[106]

Secondly, these public spaces were thought to provide long-awaited recreational alternatives that could draw the working class from their traditional leisure venues associated with intoxicating drink and lack of restraint. A report by the Council's Baths and Parks Committee expressed the view that Calthorpe Park was 'an inexpressible source of wholesome recreation and amusement; and rendered not less so, by being detached from drinking customs and the evils so lamentably in connexion with the means of pleasure in the midst of our large towns'.[107]

A third set of arguments concentrated on traditional leisure public spaces, and particularly the streets. It was reasoned that municipal parks would help reduce idling and disorderly activities such as gambling in the town's streets, especially among children and youths. One correspondent to a local newspaper supported municipal park provision as it 'would tend very much to keep the streets clear of boys who play at bandy, tip-cat, and such like games, between one and two o'clock in the day, and after seven o'clock in the evening'.[108] The Baths and Parks Committee praised the benefits resulting from the transfer of children's street games into the safe and regulated space of a park: 'The lads find a place of freedom ... playing at marbles, at football, at rounders, quoits, &c, &c, none of which could they possess in or near public thoroughfares without offending against the Bye-laws and bringing down upon them the interference of the police.'[109]

To conform to these ideals the parks were to be accessible to all working people, but on their entry inside the park railings they were to submit to surveillance, regulation and good order. The latter intentions were reflected in the park regulations of 1876, which excluded the consumption or sale of intoxicating drinks, banned any intoxicated visitors, forbade 'profane, indecent, offensive or insulting language or behaviour, gambling or soliciting alms', and prohibited all games 'except in such parts of the Parks as may be set apart for the purpose'.[110] The other objective of equal access to all working people was to be achieved by siting parks on all sides of the town. When the Baths and Parks Committee purchased land in the western suburbs, they explained they 'have for some time been prosecuting enquiries for land in the neighbourhood of Rotton Park and All Saints Wards suitable for a Public Park for that [west] side of the Borough'.[111] The eventual aim was to have a municipal park, baths and library in reasonable proximity to most working-class neighbourhoods. One Liberal Party candidate clearly expounded this objective for his own ward in 1880, arguing that it 'stands in great need of Public Baths, Free Library and Recreation Grounds; and I am in favour of immediate action being taken to secure convenient sites for the purpose'.[112] But the geography of municipal park provision, shown in Figure 2.2, was one of a dispersed scatter in the distant outer suburbs, so that the parks were a considerable distance from central and inner-suburban working-class neighbourhoods.[113]

It was not until the late 1870s that the Council began to open recreation grounds for children living more centrally, but these were only tiny islands of municipal recreational space.[114] Consequently, the parks were far less accessible public spaces than the ubiquitous courts and streets, and for many the long walk outside their own neighbourhood was a strong disincentive to more than irregular use. While these long journeys to the parks could have diminished attachments to local community, this was probably insignificant overall because of the relative infrequency of their use. This case illustrates how the geographical distribution of public facilities can significantly influence their use, but this subject has rarely been carefully assessed by social historians or even historical geographers.[115]

While the parks largely failed to transform the general frequency of usage of different public spaces, their visitor numbers are still quite impressive. During 1861 there were 415,257 visits to the one municipal park (Calthorpe Park) and by 1874 there were as many as 3,547,083 visits during the year to the increased number of four municipal parks. Between these two dates the ratio of visits over the year to total residents in the town rose sharply from 1.4 to 9.8, but viewed in relation to total activity patterns this was still quite low.[116]

Hopes that the parks would be characterised by good order when they were being used were reasonably well grounded. Disturbances or crimes reported in the local press were very rarely recorded as occurring in the parks; and the spread of organised sports in the town was also considerably hastened by the availability of park pitches. Within a year of opening the first municipal park it was reported that 'several cricket clubs have established themselves within its precincts'.[117] During 1871 almost half of the cricket matches mentioned in the *Birmingham Daily Post* (involving sixty-nine local cricket teams) were played in one of the town's three public parks.[118]

By contrast, expectations that the parks would transform people's behaviour for the time subsequent to their visit when in other public spaces were generally over-ambitious. While there was little disorder inside the municipal parks, there were problems of rowdiness and vandalism outside the entrances, such as in the Pershore Road and Bristol Road approaches to Calthorpe and Cannon Hill parks. Concern was expressed in 1863 that in the Bristol Road: 'Every day, but more especially Sundays, swarms of youths,

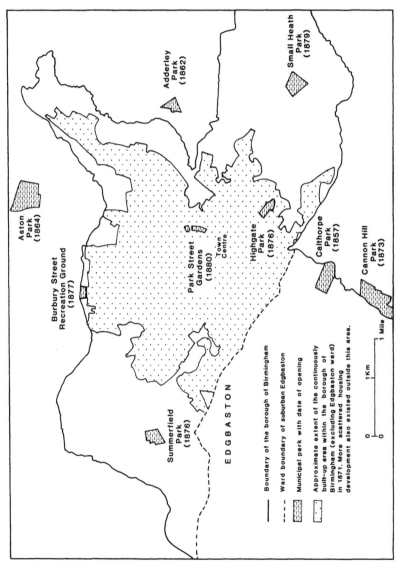

2.2 Distribution of municipal parks in Birmingham, 1840–1880
(Source: see note 113, page 140)

in gangs of four or more, infest this locality, throwing stones, climbing up and breaking the trees, stealing the flowers.'[119] One policeman also described in 1875 how 'it was an ordinary occurrence for gangs of "roughs" to promenade the Bristol Road on Sunday evenings for the "express purpose of bumping people"'.[120] A Pershore Road resident complained in 1863 how 'roughs perambulate on the Lord's day from early morn till evening' with 'the veriest scum of Birmingham in their running and dog fighting excursions on the Lord's day'.[121]

A newspaper editorial also detailed such problems around Cannon Hill park: 'There is no doubt that the great attractions of the park have led to a marked change in the former comparative quiet of the district ... on account of the exuberance of feeling on the part of the rougher element of the population.'[122] Intoxicating drink was prohibited within the park gates, but many people still combined a stroll in a park with a drink at a pub either before or after. A chairman of the Baths and Parks Committee described this connection with drink among park visitors on Sundays: 'People in large numbers go to take a walk with their families ... after walking about an hour they require some little refreshment, and they call going home at different houses that I know, and take their glass of beer.'[123] The habit of combining a stroll in a rural setting with a drink in a pub was well established even prior to the opening of municipal parks.[124] The Chief Constable explained how 'the occupations of most of our working classes are sedentary, and almost every Sunday they walk to one side of the town or the other ... and they are sure to want a glass of beer somewhere'.[125] For most working people visiting a public park was only an occasional activity and in general they could adjust to the strict park regulations during their visit without this behaviour becoming pervasive for all their activities elsewhere, including in other public spaces.[126]

By the 1870s the provision of parks by the Town Council was supported by many of Birmingham's middle class as part of a 'civic gospel' of municipal activity to secure a vision of civic advancement and common citizenship among the population. Birmingham's 'civic gospel' was promoted by a local flowering of socio-religious thought, and it encouraged elected representatives to work energetically to provide such facilities and services as parks in order to promote their perception of the common good of all the town's citizens. It held out the prospect of a new 'civic community' that would break down divisions between classes and neighbourhoods. Undoubtedly, this 'civic gospel' was prominent in civic affairs and has also featured strongly in the accounts of subsequent historians of the town.[127] The influence, however, of these ideas, and of the facilities such as parks that they helped promote, was far less pervasive than their advocates had hoped. The limited impact of the parks on working-class society has already been argued here. It is also questionable whether the abstract and diffuse notion of a common 'civic community' could profoundly affect the outlook of most working people when their daily experiences were of the immediate inequalities of work, class, neighbourhood and local community.

Birmingham and beyond

A central finding here has been that over the mid-century the continuities in the use of public space in Birmingham's working-class neighbourhoods were easily more pronounced than the changes. This conclusion does not support Daunton's view that public space was becoming 'socially neutral' or that street life was 'passive rather than participatory'.[128] To accept Daunton's ideas would be to ignore the continued importance of Birmingham's streets for active social participation among working-class neighbours and for the emergence of the social intimacies and solidarities of local communities. Similarly, local communities were important in sustaining the vitality of the street life in working-class neighbourhoods. The continuities in street society were in large measure assured because Birmingham's local communities maintained their considerable significance in promoting a sense of social relevance and personal success among the working class and in regulating working-class mores and behaviour. This second conclusion, that the town's local communities continued as important arenas for social interactions and attachments over the mid-century, challenges Standish Meacham's proposal that they did not become important until the late nineteenth century.[129]

It has been argued that the decline over the nineteenth century of some features of Birmingham's street life – notably of those involving violence and disorder, such as the wakes, cruel sports and prize fights – must be seen in the context of the many continuities in the diverse range of other social interactions that took place in the street. The preoccupation of some research on nineteenth-century society with outbursts of violence and unruliness has clearly neglected the full diversity of street life. Even the discontinuities in Birmingham's disorderly street activities were only partial. Those not involving prior planning were little changed, and there remained some people easily drawn into rowdy behaviour and others occasionally drawn into it. Discontinuity in the timing of children's street play resulting from Birmingham's educational expansion must also be seen in relation to the considerable continuities in children's activities when they did use the streets.

A great deal of research on the nineteenth century takes as its premise the notions that the working class became more respectable over the century and that this led to marked stability and even working-class parochialism in the late nineteenth century. Several contributions by social historians cited in this study are of a piece with these notions. This includes Daunton's arguments on the increasing passivity of street life and Meacham's views that working-class local community ties in towns became strong only at the end of the century. These proposals by Daunton and Meacham have been challenged in this study of Birmingham. The work of Doug Reid, which discusses Birmingham's working class in the nineteenth century more specifically, also gives great prominence both to the wide influence of a stratum among the town's working class who increasingly embraced the values of respectability and to the cumulative impact of middle-class attempts at reform and repression.[130] Clearly this essay also rejects the degree of importance that Reid attributes to such influences for Birmingham's working-class society. Instead the evidence here points to notable continuities in working-class values and behaviour, with little marked change in their overall attachments to respectability, and it warns against giving shifts in respectability an undue degree of interpretive weight.

This essay is also intended to have a general relevance for Marxist interpretations of social change. A problem for some of these interpretations is that they can rely on an overly economistic view of conflict in capitalism as being primarily, or even wholly, about struggles over work. This view overlooks a whole host of other foci of struggle that are not reducible to relations of production. These other foci include the region and the town, gender and age, recreation and religion, public space and local community.[131] A central concern of this essay has been to illustrate how the society of public spaces and of local communities are both fundamentally based on class and how these are very important foci for class formation and conflict. As such their society should receive much more sustained attention in future research.

Acknowledgements
I would like to record my gratitude to my fellow contributors for their help in improving this essay and for reviving my interest in the subject.

3

Class, culture and migrant identity: Gaelic Highlanders in urban Scotland

CHARLES W. J. WITHERS

Nineteenth-century city life was differently lived according to ties developed through daily routine – in the workplace, in social networks provided by family, friends or institutional membership and in those areas of city space in which people lived, worked and socialised. For most rural migrants city life, however and wherever lived, was additionally novel and uncertain, especially for those distinguished from the majority by facts of ethnicity, religion or language. It has been widely argued that those migrants occupied a distinct place in nineteenth-century cities and that their experience of urbanisation as a social process was reinforced by a separateness stemming from possession of a cohesive 'migrant culture' or 'way of life'; a culture that positioned them within, but apart from, other processes of urbanisation. A range of work, discussed below, has pointed to the existence and persistence of this urban migrant culture, practically organised around extended familial networks, and, in addition to common origins, based upon the shared use of certain cultural traits. Membership of particular institutions, concentration in certain ranges of employment and given districts of the city also contributed to this migrant 'identity'.

I would like to argue in this essay that uncritical use of terms such as 'migrant culture' or 'migrant identity' hides the class-based relationships operating *within* migrant groups and *between* migrants and their hosts. Further, this paper suggests that existing studies of the migrant experience in nineteenth-century British cities are open to criticism on three counts. First, they are generally based on a rather simplistic view of culture. Secondly, assessment of the migrant experience has focussed more on the 'otherness' of migrants from hosts and less on differences and distinctions within the migrant group. Thirdly, little attention has been paid to the notion of opposition and class or contradictory consciousness in the making and remaking of migrants' experience of city life.

What follows suggests that the experience by migrants of the nineteenth-century city as a social system was not dependent upon shared possession of certain cultural traits held only as distinction from others, but was, as for the city-born, framed by their relative position in given social relations of production. If this is so, the migrants' place in city life may be considered in relation to the way in which ties *both within and beyond* the migrant population facilitated or restricted entry into, and acceptance (or non-acceptance) of, the structures, institutions and values of the dominant classes in urban society. In the first part of the essay attention is paid to a range of work on urban migrant culture in the nineteenth-century city. The following section considers the matter of cultural hegemony as a means to understanding urban migrant culture. Drawing upon this review and discussion of theory, the longest section of the essay examines a particular migrant group: Gaelic Highlanders in urban Lowland Scotland. Particular attention is paid to the role of the chapel as a cultural institution and to the 1837–8 'Second report of the Commissioners of Religious Instruction' for the detailed analysis it permits of Glasgow's three Gaelic chapels. In part, this focus is the direct result of the nature of surviving source material. But in part also, this attention to migrant institutions is born of an attempt to explain social relations within migrant populations and to consider, too, both migrants' consciousness of their own identity and others' consciousness of migrants.

Migrant populations in nineteenth-century British cities

London, Liverpool, Manchester, Leeds, Bristol, York, Bradford, Wakefield, Cardiff, Huddersfield, Greenock, Dundee and Edinburgh have each provided the focus to an examination of distinct urban migrant populations in the nineteenth century.[1] The Irish figure centrally, partly because of their distinctiveness and partly because their place of origin is so easily defined. The Irish in London, we are told, possessed 'a cohesive way of life' based upon the extended family, common use of particular institutions and from concentration in certain districts. Most Irish remained 'outside and below the social organisation of the communities in which they lived'.[2] The Irish domination of particular occupations or levels of employment – 'without excelling in any branch of industry, obtaining possession of all the lowest dependents of manual labour' as the 1836 'Report on the state of the Irish poor' put it – is attested to in most studies, although one recent work would deny the claim that Irish labour was *essential* to British industrialisation.[3] If occupational concentration identified the Irish as distinct, so, too, did membership of particular institutions, their religion and patterns of residence. 'In direct contrast to the English workers...the Irish avidly pursued their religion, providing them with a source of unity and an emotional escape from the deprivation of their ghettos.'[4] When J. P. Kay wrote in the 1836 'Report on the state of the Irish poor' of the buildings erected in Little Ireland, in Irish Town and in some other of the worst parts of Manchester,[5] he was drawing attention to a phenomenon commonplace in most industrial cities, and, in the eyes of contemporary commentators, a principal cause of Irish distinctiveness.

These factors – occupational concentration, particular institutions, residential segregation, supportive kin networks and that sense of migrant identity held from within and perceived by outsiders – have been considered significant in study of other urban migrant groups.[6] In contrast to the Liverpool Irish-born, the Welsh were considered (by their hosts) 'industrious', 'enterprising', a 'steady and sober race' well integrated into Liverpool's labour force.[7] In later Victorian London, as in Liverpool, the chapel was of central significance. In contrast to the Irish, however, the Welsh in nineteenth-century London exhibited no clear patterns of residential concentration.[8] Jewish and other migrant groups likewise saw the synagogue or church as a focus.[9]

If these groups saw themselves as distinct but not apart from other city dwellers, it is also true that they were perceived by their hosts as 'something other'. Blacks in nineteenth-century British cities were viewed as an almost subhuman underclass in the context of prevalent racial theories.[10] The Irish were likewise seen as racially inferior; Celts to Anglo-Saxons. Racial notions are alone insufficient, of course, to explain that general sense of 'outcastness' with which the Irish (or others) were viewed. But in combination, the issues of class, nationality, race and religion identified the Irish in the minds of the English 'as a people set apart and everywhere rejected and despised'.[11] Jews were lauded by some commentators for their adherence to the capitalist virtues of hard work and thrift, and by others criticised, if not despised, for their control of certain industries. The Chinese were generally respected for their industriousness and their 'separateness' was tolerated.[12]

Several general points may be made as part of this survey. First, although migrant groups may have been distinct in language, religion or in some other way from others in the city – 'distinct', in this sense, by virtue of shared experiences or attributes – such populations were also internally divided along lines of class and status, across space and in terms of aspirations.[13] Simple recognition of such differences, however, without some consideration of what they signify for terms like 'migrant', 'culture' or 'host' condemns us only to a formless relativism in which migrants are perceived to become (say) less and less Irish as they drop their language and, over time, become 'assimilated' by losing their 'cultural identity'.[14]

Secondly, it may be that such an identity was maintained by some migrants as *consciously oppositional*; *because of* (as much as it was for others *despite*) their place in urban society. The status 'migrant' may then be more than *a given quality* compounded of particular attributes. It may be seen as *relational*, a class-based position in which some persons sought a solidarity with others like them (either by class or migrant 'identity') and others sought a deliberate distancing from those groups and urban social relations of which they were not part.

Thirdly, these issues are neglected in studies where migrant 'culture' or 'identity', even 'migrant community', is often treated as shared interests or practices consequent upon birthplace or a community-held ethnicity transmitted to city-born immigrant children. In one study of migrants in nineteenth-century Liverpool, for example, the term 'community' is used of migrants 'loosely to describe a group of people bound together by a particular set of common interests ... Such a group cohesion, stemming from a similar and distinctive origin, we may conveniently term 'cultural coherence'.[15] One recent overview of the Irish in nineteenth-century Britain has spoken of 'their reluctance to integrate and the host society's reluctance to accept' as functions of a 'cultural distance' between the two, a distance evident in language and institutions as 'the simple manifestations of group identity not uncommon among many immigrant groups in many lands'.[16]

Lastly, representatives of migrant culture as shared, holistic and 'traditional' and of migrants being more or less 'distanced' from those beliefs and traits characteristic of the hosts in urban society stem both from a particular view of culture[17] and from a persistent historiographical tendency to regard migrants as 'marginal men'. Thus, for Park, writing of urban migrants: 'One of the consequences of migration is to create a situation in which the same individual ... finds himself striving to live in two diverse cultural groups. The effect is to produce an unstable character – a personality type with characteristic forms of behaviour. This is the "marginal man".'[18] The idea of the urban 'marginal man' as geographically displaced but culturally distinct has had particular persistence in some North American discussions of immigrant identity and the processes of cultural assimilation,[19] though not in work on the ethnic division of labour in nineteenth- and early-twentieth-century urban America.[20]

Discussion of these questions demands attention not only to what is meant by 'migrant culture' but also to the class-based distinctions within that culture. Further, since the 'host' culture was itself riven with differences of class, status and residence, it is possible to suggest that cultural processes of urbanisation were more complex than is often supposed in treating acculturation of urban migrants as a one-way process as incomers adopt the mores and behaviour of the majority. Migrants made their culture themselves by adapting the beliefs and practices they brought with them, but there were constraints imposed on that making by circumstances they found in the cities.

Class, hegemony and migrant culture

Consideration of culture as consensual and universally shared ignores explanations of cultural characteristics and fails to consider culture as a process of social interaction.[21] Indeed, attempts to use anthropological concepts of culture have little value in examining class-based societies.

As noted above in the Introduction (see also Chapter 1), the concept of class is dynamic and relational, based on the social organisation of economic relations. But any discussion of culture and migrant culture that sees culture only as traditional beliefs or a shared system of values will lack that relational sense, deny the validity of class consciousness as a source of cultural identity and limit understanding of the socially constituted relations between and within groups of people: 'In class society culture is the product of class experience. The common-sense reflections of each class upon its material experience is part of its struggle with other classes, each attempting to impose what it sees as the universal validity of that experience.'[22]

In the context of migrant populations, then, processes of socialisation are neither with and through one culture or class nor with the city as a whole. Culture is not a rigid social form. It functions 'as a dynamic mediating mechanism – helping its adherent adapt to rapid social and environmental change'.[23] 'Migrant culture' as a persistent 'identity' may thus be considered both as a response to the prevailing hegemonic attitudes of the upper classes and to the actions of migration.

The notion of hegemony as a means to understanding relations between social classes – and, more especially, the cultural dominance of one class and the means by which that class maintains its dominance – is associated with the work of Gramsci. Gramsci distinguished between 'rule', expressed in overtly political forms and at times of origin by direct coercion or military intervention, and 'hegemony', the exertion of consensual relationships through intellectual, moral, economic and social forces, expressed and manipulated through particular institutions and practically organised by specific dominant values.[24] Hegemony is not an absolute condition but a *process* of class domination.

It is a whole body of practices and expectations, over the whole of living: our senses and assignments of energy, our shaping perception of ourselves and our world. It is a lived system of meanings and values – constitutive and constituting – which as they are experienced as practices appear as reciprocally confirming. It is, that is to say, in the strongest sense a 'culture' which has to be seen as the lived dominance and subordination of particular classes.[25]

In Gramsci's terms, hegemonic domination within civil society depends principally upon the role of intellectuals, institutions and language.

Gramsci distinguishes between 'organic' intellectuals and 'traditional' intellectuals. The first are tied to the ruling class; they represent it and give it an awareness of its own function across all classes. The second comprises both individuals regarded as intellectuals within the *prevailing* socio-economic formation and the vestiges of organic intellectuals from *previous* social formations (such as ministers). Institutions mean not only the church or educational authorities but also the family, the workplace and the clan.[26] Language has two meanings in this context. In the simplest sense, language arises from the need to communicate. But it also gets its potency through particular social relations. Language as the expression and interpretation of material and intellectual production is crucial to the maintenance of class-based cultural hegemony: 'Language, like capital, is an instrument of domination, a carrier of cultural power.'[27] For Dupré: 'The class that rules succeeds in presenting its particular use of language as the only correct one...[Language] does not "reflect" reality, it *expresses* and *represents* it.'[28] Language is thus at one level a cultural 'given' – an attribute possessed, or not, by urban migrants in ways that *may* act to maintain 'distinctiveness' but not necessarily to determine 'separateness' from others in the city. At another level, it is an incorporative element of material and social relations through which cultural hegemony is established and maintained.[29]

Hegemony as a process and system of dominant values is neither held absolutely by all nor without internal contradictions.

We have to emphasise that hegemony is not singular; indeed that its own internal structures are highly complex, and have continually to be renewed, recreated and defended; and by the same token, that they can be continually challenged and in certain respects modified. It is also continually resisted, limited, altered, challenged by pressures not all its own.[30]

In the context of opposition to dominant hegemony, we may distinguish between alternative hegemony and counter-hegemony.[31] By alternative hegemony is meant the emergence and consolidation of a new hegemonic culture, associated with the rise to power of a new class. Alternative hegemonies remain, however, unrealised: for Thompson at least, the idea is 'inapplicable to a subordinate class which by the nature of its situation cannot dominate the ethos of a society'.[32] Counter-hegemony is an opposition to the prevailing hegemony that may or may not be motivated by an articulated class consciousness but is nevertheless characterised by attempts to limit or constrain dominant hegemony. Counter-hegemony may take several forms – the continued use of a 'minority' language, establishment of particular clubs or societies, or more politicised forms such as food riots, machine-breaking, land wars and refusal to pay rent.

We may further consider the distinction between 'residual' and 'emergent' oppositional culture. By 'residual' is meant those meanings, values and experiences that cannot be expressed in terms of the relations and classes of the dominant culture yet remain lived and practised as part of the residue of some previous social formation. Residual notions – language, house styles and so on – may, in time, be incorporated into, and survive within, the dominant culture.

'Emergent' means not only new values and practices being created within society but also how dominant culture reacts to those new forms. Emergent cultural forms may be simply alternative without being oppositional. But as the area of control of those emergent forms extends, what was once casual disregard for dominant values may become open dissent; and the forms and institutions of dominant culture may need to defend and modify, to incorporate oppositional claims.

The notion of hegemony has been used by several geographers. Billinge's discussions of early-nineteenth-century Manchester notes that hegemony allows 'an entirely new perspective from which to view the progress of acculturation and the crucial importance of culture in shaping social ideas and social relations'. It does so because culture is no longer seen as an assemblage of characteristics and attributes simply dependent upon more 'basic' economic and political forces, but is, rather, 'amongst the most basic of processes involved in the actual formation of relations between social classes'.[33] Drawing upon the work of Poulantzas and Gray, Billinge additionally notes that the power bloc in nineteenth-century society was 'far from monolithic': 'The power bloc – a collection of cooperative interest groups which together constitute all social power – is seen not as synonymous with the hegemonic group alone but more subtly as a shifting alliance between *politically* dominant, *culturally* dominant and *ideologically* dominant social sub-groups which may be drawn from different social classes.'[34]

Mouffé has likewise shown how hegemony is not simply a class subjugation but a process by which the dominant class articulates the interest of other social groups as its own means of ideological struggle.[35] Agnew and others have drawn upon the notion of hegemony in explaining the city in cultural context: 'To put the city in cultural context is to view it as the product of both hegemonic and subordinate cultures and, at the same time, as the site for their production.' Whilst recognising, however, the nuances attached to the term 'culture' and rejecting the relativism of schemes that typify urban and rural culture as opposing ends of a continuum, Agnew et al. embrace the term 'cultural context' rather than 'culture', 'because we do not want to be tied to the concept of culture as national or hegemonic. The term we prefer should suggest a nesting of contexts, from class and ethnic to national and global, by which specific cultures are defined and relate to one another.'[36] This is borne of a claim that 'most recent usage [of culture] has tended toward a consensus in taking national societies as the units to which the concept of culture is most appropriately applied'.[37] It is a claim that finds support in a recent examination of English state formation as, broadly, a 'cultural revolution' in which the process of 'moral regulation' as a means of making particular and historical forms of social order seem natural and permanent approximates in general outline to notions of hegemony and counter-hegemony.[38]

Although there are difficulties with the use of cultural hegemony – not least in understanding what one historian has called 'the apparent contradiction between the power wielded by dominant groups and the relative cultural autonomy of subordinate groups when they victimise'[39] – it is suggested here that the notion of hegemony has value in explaining the forms of migrant identity over city-space and across different social classes, the role of migrant institutions, the readjusting accommodation of the dominant classes to the desires of subordinate groups and the way in which some migrants, in developing social relations with those of dominant class, may have shared a kind of half-conscious complicity in the subordination of their own culture. Notions of hegemony make it possible to consider migrant culture as both part of dominant class structure in the city and a 'residual' element possessing meanings, values and experiences that cannot be expressed in terms of dominant class interests yet remain lived and practised.

Migrant culture, however differently lived and expressed according to the social class of migrants and the alliances sought within and beyond their class and culture, took the forms that it did in particular cities for a variety of reasons: partly as reflection, in residual form, of economic formations and social relationships migrants brought with them to the city; partly in consequence of migrants' (largely subordinate) position within urban society; and partly as a consciously oppositional strategy on the part of some migrants. Since hegemony as a process of class-based cultural dominance may be both internally fragmented and externally challenged, this model allows for the possibility that some migrants could seek accommodation with those of unlike culture without being represented as 'less Irish' or apart from migrant culture altogether. And, if some migrants effected compromise with the dominant 'power bloc' in urban society, it is clear, too, that for others, the sense and strength of migrant identity, the existence of specific institutions or the persistence of residential *locale* might be considered as the conscious assertion of counter-hegemonic interests, themselves structured on class lines, and not merely the maintenance of cultural traits stemming from shared belief or the fact of common birthplace.

With these ideas in mind, the remainder of this essay examines Gaelic Highland migrants in nineteenth-century Glasgow and the role of chapels there as cultural institutions.

Only from 1851 is it possible to assess numbers of migrant Highlanders in Lowland towns and know their place of origin in detail. Table 3.1 shows the number of Highland-born in the principal Scottish cities in 1851. Highlanders migrated both in consequence of socio-economic changes in Highland rural life and in response to the employment opportunities in the urban lowlands.[40] The construction of Gaelic chapels in those cities testifies to the presence of a permanent Highland migrant population who wished to worship in Gaelic even when English was used in other social domains. Gaelic chapels were built in Edinburgh in 1769, in Perth in 1787, in Aberdeen and Dundee in 1781 and 1791 respectively and in the western towns of Greenock and Paisley in 1792 and 1793. Glasgow had two Gaelic chapels, in Ingram Street, built in 1770, and in Duke Street, built in 1798. Chapels were built by subscription

Table 3.1. *Birthplace, by county, of the Highland-born population of selected Lowland towns, 1851*

Highland county	Lowland towns* (Proportion Highland-born as %, by county, of Highland total in indicated town)							
	Aberdeen	Dundee	Edinburgh	Glasgow	Greenock	Paisley	Perth	Stirling
Argyll	2.3	6.7	15.59	70.69	78.67	80.0	2.3	26.0
Bute	0.2	0.6	1.10	6.50	9.30	3.8	0.7	1.9
Inverness	34.7	27.2	26.60	13.64	9.18	11.8	12.3	19.5
Ross and Cromarty	40.4	15.8	25.30	4.46	2.09	1.8	3.1	7.9
Sutherland	15.8	4.8	9.18	2.58	0.23	0.4	1.6	5.6
Highland Caithness	4.3	1.4	4.79	0.23	0.07	0.3	—	—
Highland Moray	1.5	0.1	0.28	0.08	—	—	0.3	—
Highland Perth	0.6	43.0	17.20	1.68	0.44	—	79.6	39.0
Highland Nairn	0.2	0.4	0.16	0.06	0.02	1.0	0.1	—
Total number of Highland-born	1,762	809	4,303	14,959	4,243	1,584	1,200	215
Highland population as % total urban population	2.44	1.02	2.68	4.54	11.33	3.30	5.03	1.67

* All of the Lowland towns indicated are based on the parliamentary burghs.

(*Source: Census of Scotland 1851, census enumerators' books*)

from amongst the Highland-born population in each town and managed by a committee on behalf of migrants as a whole. Many members of these committees were better-off Highlanders who saw the chapel as a means of using positions of authority within migrant populations to facilitate their own admission into the power structure of the English-speaking host society.[41]

In the 1830s Glasgow had three Gaelic chapels: at Duke Street, administered by the Rev. Lewis Rose; at Ingram Street (known as St Columba's parish although it was located in St George's parish), the charge of the Rev. Dr Norman McLeod; and at Hope Street (also known as West Gaelic Church), under the Rev. Hector McNeill. Each was a *quoad sacra* parish church, that is without formal parochial limits 'in consequence of the scattered situation of the population'. These men and others from the Highland congregations in Glasgow drew up a 'Gaelic census' of Glasgow in 1835-6, which was presented to the Church Commissioners. This, together with the Commissioners' report on the Gaelic chapels, provides the basis to what follows.

Gaelic chapels, institutional hegemony and urban Highland culture: the case of Glasgow

The original census returns do not survive. We are told only of volumes that contained:

The names of all the individuals of the adult [Highland] population in Glasgow, the parishes in the Highlands where they were born, the trade and occupation which they follow in Glasgow, the street and the lands in which they reside, the number of their families under and above ten years of age, the churches which they profess to attend, according to their own statements, and also the number of seats which they occupy.

The result of their enquiry was that:

There are 22,509 native Highlanders in Glasgow, including all descriptions of persons. Of that number 5,336 are under 10 years of age; 2,529 have seats in churches where the English is preached. 3,102 attend the three Gaelic churches, and 11,522 adult persons have no seat in any church, and these make up the entire number stated. It will be found from these lists, that 1,529 who attend the churches in which English is preached, are of the wealthier portion of the community as may be expected. It will be seen that they are merchants, and people in the respectable walks of society, who have prospered in life and trade, and that the greater proportion of the 11,522 having no seats, will be found to be labourers at public works, employed as operatives and labourers. This is the result of the census we have taken.[42]

The difference in total for the Highland migrant population in Glasgow of 22,509 persons (in 1836) and 14,959 persons in 1851 (Table 3.1) is easily accounted for and hints at a crucial distinction underlying migrant identity here. The 1836 figures enumerated all those 'resident within 3 or 4 miles of the city' *who were able to speak the Gaelic language*. Ability in Gaelic, not

Table 3.2. *Gaelic speaking in urban Lowland Scotland, 1891: the examples of Aberdeen, Dundee, Perth and Stirling*

	Aberdeen	Dundee	Perth	Stirling
Total Highland-born population	1,257	1,277	1,117	376
Total Gaelic-speaking population (from enumerators' books)	660	760	815	482
Total Highland-born Gaelic-speaking population, and as %	514	555	595	229
of total Highland-born	40.89	43.46	53.26	60.9
Total non-Highland-born Gaelic-speaking population, and as %	146	205	220	253
of total Gaelic-speaking population	22.12	26.97	26.99	52.48
Male Gaelic speakers as % of total male Highland-born	44.14	45.17	53.09	58.06
Female Gaelic speakers as % of total female Highland-born	37.69	41.88	54.36	62.89

(Source: Census of Scotland 1891, census enumerators' books)

Highland birth, qualified one as Highland: 'Many of the persons included in the above 22,509 were born in Glasgow, and most of them could speak English, but all of them were able to speak Gaelic. Many who had been born in the Highlands, spoke English imperfectly, some of them not at all.' Although the takers of the religious census confined themselves 'to those who are *bona fide* native Highlanders, of whom many attend English, as well as Gaelic Churches', they included Glasgow-born Gaelic speakers in the Highland population because 'they have been educated at firesides where the Gaelic only is spoken from morning to nights; and some of them are the best speakers of Gaelic, and the most zealous Highlanders'. The 1836 enumerators omitted English speakers: 'They were directed to exclude every person born of a Highland family in the Lowlands, and speaking English. I could specify [testified Lachlan McLean on 26 March 1836] one or two families, out of which they have taken one and left five.' Superintendents corrected returns when, as McLean said, 'there were [Gaelic-speaking] people left in particular instances that I knew'.[43]

These linguistic distinctions raise important questions about language in relation to migrant identity. Unfortunately, it is only possible to know in detail the numbers speaking Gaelic in Lowland cities from the enumerators' books of 1891.[44] Table 3.2 shows the numerical strength of Gaelic in four other Lowland cities in 1891. Several points are worth noting. First, between only 40 and 60 per cent of the Highland-born population spoke Gaelic. Secondly, no clear distinction emerged as to the use of Gaelic by sex: in

Aberdeen and Dundee more Highland-born males than females spoke the language; in Perth and Stirling the case was reversed. Thirdly, about a quarter (in Stirling over a half) of the Gaelic-speaking population were non-Highland by birth. This in turn represents three factors: most importantly, ability in Gaelic of Lowland-born children of Highland-born Gaelic speakers; Gaelic speakers from overseas, presumably descendants of Highland emigrants; and numbers of Irish-born registering ability to speak Gaelic, for whom ability in *Scottish* Gaelic as opposed to Irish must be questioned.

Within the Highland-born Gaelic-speaking population in these towns, the numbers speaking Gaelic in different age cohorts suggest that Gaelic was used more by persons within the working-age range – from about 20–24 to 60–64 years of age – than by persons younger or older. At the family level, the most commonly occurring distinction was the listing as Gaelic speaking of one or both Highland-born parents, with non-Gaelic-speaking children. This part-Highland part-Gaelic family was more common than either entirely Gaelic-speaking wholly Highland-born families or families and households where the eldest child and/or lodgers spoke Gaelic together with one or both parents.[45]

These data may not be directly transferable to the case of Glasgow in the 1830s. But it is clear that Gaelic and English were not only both known and differently used within the Highland population, but that they were known and used in many ways dependent upon age, social class and deliberate preference.

In all three chapels both Gaelic and English services were held. In Duke Street: 'The attendance is smallest in the forenoon when Gaelic is spoken, as children who understand it imperfectly attend the afternoon service only, which is in English.' Elsewhere, Gaelic services were not as well attended 'owing to the circumstance that the parents themselves prefer the Gaelic service, and their children do not'. The Rev. McColl, a Gaelic missionary employed in St Columba's parish, was asked if 'children born in Glasgow of Highland parents, and bred up in the knowledge of Gaelic and English, would...be inclined, in after life, to go to a Gaelic church'. His answer – 'Yes, because a Gaelic sermon will make a far greater impression on those who understand the Gaelic language' – points in one sense to a deliberate and widespread preference for Gaelic in religious worship. In another sense, however, this is contradicted by those who observed how Lowland-born children of Highland-born Gaelic-speaking parents were often only partly able in Gaelic, even when that language was used in daily conversation by their parents at home.

Such generational shift in language ability was common in other nineteenth-century Gaelic populations and earlier and occurs widely today in other contexts.[46]

In Glasgow the adoption of English by Highland-born children and children of Highland-born (many of whom were Glasgow-born part-Gaelic speakers) was influenced by schooling policies operated through the Gaelic chapel. Since its foundation in 1727, the Glasgow Highland Society had run schools in Glasgow 'for the exclusive benefit of the children of Highlanders'. In 1836 attendance was 500 boys and 200 girls. Seats were reserved at the afternoon English service in Ingram Street chapel for between 80 and 100 boys. The Society also funded the purchase of English-language school-books for Highland children.[47] Together with the chapel management and through the use of chapel facilities, the intent of such education was, from the outset, to anglicise the Highland community as a means to their broader education and social improvement.

The boys after a certain number of years at school are apprenticed to trades, and have the benefit of instruction at night in the schools; and the girls are taught English reading, writing, arithmetic – the simple branches of it – the branches of female industry – they are prepared for being respectable servant-maids. A certain number of the boys and girls receive clothes. The education is given in English.
We accommodate from 80 to 100 of the older boys, of those who are furthest advanced, in my church [Ingram Street] in the afternoon, and these seats are occupied by poor Highlanders in the forenoon.[48]

English was both the symbol of and means to authority in nineteenth-century urban society. Gaelic, however much it was preferred in church worship and used in other domains, was regarded (by non-Gaelic speakers) as backward, its users uncultured. The Glasgow Gaelic Highlander was regarded as linguistically and socially uncultivated because he was, both occupationally and in terms of perceived class consciousness, lower class. Language and social class were mutually supportive parts to migrant identity.

The truth is – and intelligent Highlanders will be the first to acknowledge ... – that the labouring population in the Highlands, regarded as a whole, are behind their neighbours in other parts of the country in the march of civilisation. They are deficient in the knowledge of letters, they are still more deficient in that practical education of civilised life, which results from the mutual attrition and jostling of parties endeavouring to outstrip each other in the acquisitions of industry.[49]

The urban Highlander was viewed as 'behind ... in the march of civilisation' in the same way as contemporary ethnological theory saw the migrant Irish as 'outside and below'. In a related way, and in combination with persistent images of the Gaelic Highlander as almost physiologically ill-suited to the work ethics and enterprise of the superior (Saxon) Lowlander,[50] this meant that one part of the identity attaching to the urban Highland migrant was determined *prior* to arrival (for parallels, see Chapter 4). Migrant identity was framed not from within by urban migrants themselves, but, upon arrival, from 'above and outside' in the contemporary consciousness of the urban middle class.

Distinctions of language and class were also the basis for differing migrant identities. In each chapel Gaelic was much preferred in religious worship by 'the labouring classes', English by 'the wealthier section of the population'. This point was stressed again and again. There was what McLeod considered 'a great desire among that class [the 'working poor'] to have service in their native language'.

This choice was partly born of a belief in the oratorical rhetoric and symbolic power of Gaelic sermons over English: 'It is a feeling of nature, which, like the mountain torrent, not only moves, but sweeps everything before it'. In contrast, in English preaching, the sermon is made up of 100 different languages, 'which it takes a person always eight years' hard study to make himself master of'.[51] It was partly also the result of relative inability in English amongst Glasgow's Highland working class. Lachlan McLean noted that 'a great proportion of them can do little or no business whatever in the English language – and in corroboration of this fact, I have myself got Highlanders into situations, warehouses, and shops, the masters of whom took them chiefly, if not altogether, because they could talk Gaelic, to those who could not talk English to their masters'.[52] The role of Gaelic Highlanders like McLean and chapel ministers as cultural 'brokers' in this way is both widespread and vital.

What is crucially important, too, is the fact that Gaelic services for the Highland poor could only be given *at all* because of the provision of English services for the better-off Highlander. This is not the paradox it seems. The answer lies in seat rents; charges levied by church managers to permit an individual a position in church. Seat rents, in Glasgow and in other nineteenth-century towns, segregated congregations by class.[53] In Glasgow's Gaelic chapels, seat rents paid by the wealthy Highlanders to attend English services were much higher than those paid by labouring Highlanders attending Gaelic services. In some cases, seat rents for Gaelic services were not paid at all because of the poverty of the Highlanders. Many were said to be 'deterred from coming to the public worship by want of proper clothing' and because, in McLeod's words, Highlanders were 'extremely averse to the system of seat-rents ... they never heard of it till they came to Glasgow'.[54]

The poverty of those Glaswegian Highlanders who preferred Gaelic services (a poverty evident in inability to pay seat rents) ensured the continued existence of English services within Gaelic chapels. The position in urban social relations of this section of Highland migrants was, simultaneously, the basis for a given identity in the city and the means to their linguistic and cultural transformation, no matter how much they continued to speak Gaelic at home, at work or preferred it in church worship (and despite the fact they may have shared the same birthplace with English-speaking Highlanders). However much these Gaelic Highlanders might have wanted to secure an identity in opposition to the claims of others through a Gaelic-only chapel, their numbers, their scattered distribution throughout the city and principally their poverty determined otherwise. For McLeod:

The chief obstacle to the prosperity of the Gaelic churches is the practice of having sermons in the two languages in the same day, which he thinks to be necessary, as those who understand Gaelic only are of the working classes, and unable to maintain a church exclusively for their own use; and that therefore an English service must be performed once a day, to induce the wealthier Highlanders who understand the English to take the seats in the church ... If a church exclusively for Gaelic were set on foot, there would be a lower standard of fashion as to dress, and an attendance of the working classes in their working dresses ... But without an endowment ... the latter could not be upheld, as, partly from their extreme aversion to the system of seat-rents, to which they had not been accustomed before coming to Glasgow, no adequate revenue from that source could be expected.[55]

In Duke Street Gaelic chapel, the question of seat rents was both a matter of poverty and an aspect of Highlanders' 'residual' culture.

This church has been of late years upheld chiefly by the attachment and liberality of a very few individuals ... We have enough of accommodation for the paying Gaelic population of Glasgow, but there is a subordinate Gaelic population, either unable to pay, or that, owing to various circumstances, have been so much out of the habit of attending public worship that they are now too careless about it ... The Highlanders have not, in their earlier days, been accustomed to pay any seat-rents, and that the very idea of paying seat-rents is abhorrent to their nature, and the privileges they formerly enjoyed.[56]

In the West Gaelic church, poverty and aversion to seat rents meant that whole Highland families could not attend Gaelic services: 'They do not take a sufficient number of sittings for the accommodation of their families, seldom renting more than one for the husband and wife, and several grown up children of ten or twelve years of age'. In Ingram Street, many borrowed clothes to come to church 'because a Highlander supposes it to be the greatest degradation possible to be considered such a character as not to attend divine worship'.

Several points may be made in review here. Attempts to establish a Gaelic-only church to provide for Highland migrants whose identity crucially depended on worship in that language were doomed. Gaelic-only services were an impossibility given the position of possible participants in the urban environment – their poverty, their scattered residential patterns and the attitudes they brought with them to the city – adherence to Gaelic and an aversion to seat rents. The last was a 'residual' aversion overcome, however, by Highlanders wealthy enough to do so. Such persons could use that wealth in ways that determined that better-off Highlanders, Gaelic speakers amongst them, adopted English through institutions which, at one and the same moment, sought to maintain migrant identity through provision of Gaelic services for persons of like birthplace but different class. There is some evidence to suggest that differences within the town's Highland population stemmed from distinctions in origin. Duke Street congregation was held to be chiefly 'North Country Highlanders'. McLeod spoke of 'the great diversity of tongues and brogues among Highlanders from Sutherlandshire, Inverness-shire and Argyllshire, who speak in their different dialects', noting how, for reasons of familiarity, many of Glasgow's Highlanders 'generally prefer a man from the west country'.

The wealthier part of the population, in general, are from the districts of the Highlands nearest from the low country, where they have the advantages of schools, with the exception of Ross-shire, at an earlier period than other districts of the Highlands. The poorer class, and by far the greater proportion of the Highland population, are from the islands, and the western coast of Argyllshire, and Inverness-shire.[57]

Table 3.1 would support the last part of this statement. McLeod elsewhere confirms his belief in the first part. In testifying to an 1841 Parliamentary committee, he noted: 'There are very few works in Glasgow in which Highlanders are not found; the classes who have received the benefit of education are employed in warehouses; they are employed as clerks, and so on; the other classes, the uneducated, are employed at various public works.' Highland labourers were 'invariably from the districts where education is lowest, whereas in the localities where education has been best cultivated they are found of a higher grade of society, and rise above the character of labourers'.[58] Unfortunately, there is no way either of knowing the educational attainments of individuals migrating or of relating level of education to type of urban employment.

There is, nevertheless, good reason to suggest that Glasgow's Highland population was structured on grounds of local loyalties as well as on grounds of class and language and was influenced, too, by the earlier presence in that town of kinsmen or persons of like background for whom shared identity extended across (as it also lay within) divisions of class.

I think [wrote McLeod] Highlanders find it more easy to get respectable employment than the Irish; the Highlanders have many friends in Glasgow to whom they apply; there are very few days in which we do not receive letters of recommendation on behalf of poor Highlanders coming to Glasgow; they come with letters of recommendation to countrymen and clansmen who are in comfortable circumstances; we are very clannish; and those who come from one Island do it for the men from that Island who have to get employment – the MacDonalds for the MacDonalds and the McLeods for the McLeods and so on, so that they find very little difficulty in getting work.[59]

There were thirty-eight Highland societies of various kinds operating in nineteenth-century Glasgow.[60] Many had previously defined specific loyalties. The Glasgow Argyllshire Society, set up in 1851, offered 'pecuniary relief to its Ordinary Members and Their Families, and other persons connected with the County of Argyll by birth…[and] the promotion and extension of education among poor children connected with Argyllshire, whether resident there or in Glasgow'.[61] The first object of the Uist and Barra Association (founded 1890) was to 'afford assistance to its own Members' and to obtain situations for persons who have come from North Uist, Benbecula, South Uist and Barra in search of employment.[62]

In terms, then, of an internally motivated identity – migrant consciousness of their own identity and the means to make and remake that identity – the label 'urban Highlander' was underlain by local loyalties to place and clan, however much it was also externally framed in class terms in the opposing consciousness of the non-Highland urban middle class.

It is difficult to know how this identity was influenced, either from within the Highland population or in the minds of other Glaswegians, by the residential patterns of the Highland-born. The evidence in the Commissioners' report is partial. The congregation of Ingram Street chapel, although 'much dispersed' throughout the city (thus occasioning problems in pastoral superintendence) was also said by McLeod to be congregated near the Broomielaw and in the Bridgeton and Anderston areas of town. This information on relative concentration has been supported by other research.[63] But in the absence of nominative congregation lists, it is difficult to know for certain either if Glasgow's Highland-born Gaelic *chapel-going* population was proportionately residentially concentrated by class (or other division) in given parts of the town or if residence in particular districts determined attendance to one chapel in preference to the others.

From records relating to Highland-born paupers in Glasgow in the later nineteenth century,[64] it is possible to hint at the extent of residential mobility of some Highland-born poor. Table 3.3 shows the number of listed residences for a sample of 157 Highland-born paupers in Glasgow's City parish from 1851–97. Although the figures sampled are small, this evidence suggests that there was no direct correlation between length of residence in Glasgow and frequency of residential change for the Highland-born poor. Very few of the more recently arrived migrants moved more than twice. In contrast, nearly 50 per cent of those resident longer than ten years had moved on six or more occasions. Many moves were short distance, some literally 'across the stair' within the same tenement.

Whilst it may be suggested that many Highland-born poor were relatively circumscribed in their district of residence even though they moved frequently, it is another thing still to claim from this evidence that such 'segregation' underlay or in any way acted upon the migrant's sense of identity, not least given the other ways in which we have seen that identity to have been structured.

Table 3.3. *Residential mobility and length of residence in Glasgow of a sample of Highland-born residents in City Parish, Glasgow, 1851–1897*

Length of residence in Glasgow (years)	Number of listed residences in Glasgow since arrival (figures in parenthesis = % of total number of moves made by persons in given time period)						Total number of moves made, by period
	1	2	3	4	5	6	
–1	9 (41.0)	1 (4.3)	1 (4.1)	—	—	1 (2.1)	12
1–2	—	4 (17.4)	—	—	—	—	4
2–5	1 (4.5)	2 (8.6)	9 (37.5)	2 (8.7)	7 (38.8)	7 (14.9)	28
5–10	4 (18.0)	7 (30.4)	6 (25.0)	12 (52.2)	5 (27.9)	16 (34.1)	50
10 years +	8 (36.5)	9 (39.3)	8 (33.4)	9 (39.1)	6 (33.3)	23 (48.9)	63
Totals	22 (100.0)	23 (100.0)	24 (100.0)	23 (100.0)	18 (100.0)	47 (100.0)	157

(*Source*: Glasgow City Parish, General Registers of Poor and Applications for Relief, Strathclyde Regional Archives, DHEW (*1851–1855*) (*1855–1862*) (*1864–1871*) (*1874–1883*) (*1884–1889*) (*1894–1897*))

Even assuming that residential propinquity with others of like place of birth was a measure of community spirit or helped form a specifically migrant identity, the evidence for Glasgow's Highland-born poor would suggest that residential mobility was high, commonplace with a frequency of move for some individuals of two or even three times a year. It would suggest, too, that even if specifically Highland parts of given Lowland towns existed to the same degree of concentration as has been argued for the urban Irish elsewhere, such spatial identities as existed were underlain in complex ways by other social divisions more meaningful to the making of migrant identity. Smith has argued that 'Glasgow was not a city of ethnic neighbourhoods, either of the Irish or the Highlanders.'[65] It may be, then, that a lower-class Highland identity persisted as a result of relative residential dispersal given their irregular attendance at chapel and the difficulties faced by attending clergy.

If, using the census, we may distinguish between wholly Highland urban families, where both parents and all recorded children were Highland-born, and 'part-Highland families', where both parents were of Highland birth but none of the children were Highland or where one parent and the eldest child were Highland-born, the great majority of Highlanders in the urban Lowlands fell in the latter group.[66] In Barony parish, Glasgow, in 1851 there were 4,491 Highland-born persons (30 per cent of the total Highland-born population in the city). Only a little over one-fifth (20.8 per cent) of the Highland-born population in the parish were wholly Highland families; 41.7 per cent were part-Highland families where only the parents were Highland-born; and 37.5 per cent were part-Highland families with only one Highland parent.[67]

Whilst some Highland families and households did offer assistance to fellow migrants, kin or not, marriage with non-Highlanders was more common than intermarriage between Highland-born partners. Certainly, there is no evidence to support the idea of urban Highland 'marriage communities' in the way suggested of the Irish in some nineteenth-century cities.[68]

Class distinctions were apparent in the chapels, not only in regard to the inclusion of afternoon English services for the wealthy English-speaking section of Glasgow's Highland-born migrants and in levels demanded for seat rents, but also in policies of management. Chapel managers were appointed for terms of five years and, once elected, formed a virtually dictatorial élite within the congregation.

During their tenure of office, they fill all vacancies in their number, without any reference to the seat-holders. They may burden the property to any extent they choose ... No qualification is necessary for becoming a manager of this church, such as being a member of the congregation, or the possession of a sitting, and the attending on the divine worship for a certain period previous to the appointment. The managers are the sole administrators of the funds, without any control or inspection over their proceedings by the contributors.[69]

Alexander Munro, one of the managers of Duke Street Gaelic Church, referred to men of extreme religious opinion 'and feelings adverse to the prosperity of this church' who had insinuated themselves into the management 'for the promotion of party or electioneering purposes, and retained the office almost a lifetime, often to the annoyance of both minister and congregation...These men, by the mere possession of a sitting and a good address, were enabled to creep into the self-elected managements of several chapels.'

In such cases, this conflict within migrant institutions was opposed by men like McLeod, acting as cultural 'brokers' for the groups they represented.

A number of independent and public-spirited Highlanders, viewing the conduct of those men with indignation, stepped forward, and backed the skeleton of a congregation that remained to the house, and, at considerable pecuniary sacrifice, secured the ministrations of an eminent minister from a Highland parish; and the consequence is, that there are above 1,000 sittings let already to Highlanders, while at the same time, the church is quite full every Sabbath.[70]

In this sense, too, Gaelic urban migrant culture was reinforced from within through the congregations it served and the wider Highland culture of which it was part, despite those divisions of class, language and local loyalty on which it was also structured.

Hegemony, culture and migrant identity

Gaelic chapels played a key role in framing the identity of Glasgow's Highland migrants. But they did not do so in the same way for all 'Highland-born'. Indeed, given the above evidence on the divisions within Glasgow's Highland population, it is possible to claim that the notion of a *single* migrant identity is unsustainable.

Lower-class Gaelic-speaking Highlanders sought an identity through religious worship in a language many of them spoke at home and at work. Their identity, their own consciousness of their position in society, was influenced not alone by the fact of Highland birth but by their subordinate position in urban society. Wealthier Highlanders and those Highland-born who spoke English were no less Highland despite attending English services. Yet the identity of these people was not framed in the consciousness of the urban middle classes in the same way as the lower-class Gaelic Highlander whose language and class, not his place of birth, set him apart from natives of the city and other Highland-born alike. And within the Highland-born Gaelic population, identity was being continually remade by the relative inability of children to speak and maintain Gaelic.

For the Highlander (Gaelic speaker or not) whose wealth allowed him to pay seat rents to hear English services in a Gaelic chapel, such language shift was a means to incorporation into the hegemonic structures of the ruling groups, within the migrant population and in society at large. For the Gaelic Highlander in particular, the full realisation of an identity in conscious opposition to the English-speaking urban power bloc was hindered by several factors. The first was language loss over time amongst second-generation Gaelic-speaking Highland-born. The second was the inability through poverty of the great proportion of Gaelic Highland-born to attend Gaelic services regularly. And thirdly, even if Gaelic Highlanders had been able to do so, funds raised by seat rents were insufficient to sustain Gaelic-only services. Lastly, the generally scattered residential distribution of Highlanders detracted from the formation of an articulated counter-hegemony amongst working-class Gaelic Highlanders. In one sense, then, Highland identity persisted amongst the poorest urban Highland-born because of their total immersion in the culture of poverty.

For these reasons, contradictory consciousness amongst Glasgow's working-class Gaelic Highlanders existed only as alternative, not as articulated, counter-hegemony. It is clear also that Highland identity, however evident by class or language within the Highland-born population, was structured in other ways. The role of migrant institutions is of particular importance. Highland secular institutions cooperated in what, often simultaneously, were general philanthropic functions and means to specific class-based cultural transformations. The Glasgow Highland Society used the profits from their Black Bull Hotel to fund its schemes for the apprenticeship and education in English of Highland children: 'Generations of unkempt young Highlanders have been made intelligent and attractive citizens, on the profits of hundreds of pipes of claret, port, and sherry, thousands of puncheons of whisky, and oceans of Glasgow punch.'[71] Other educational bodies like the Glasgow Auxiliary Society for the Support of Gaelic Schools (founded in 1812) – which continually entreated its committees of management and supporting constituency 'not to forget your less-favoured countrymen' – point to a class-based system of support and general migrant identity that was at once local, national and international in extent.[72] As Smith has noted elsewhere, class conflicts, or at least issues of contradictory consciousness, were often reduced to matters of compromise and support resolved through specific formal bodies.[73] Highland migrant institutions were no exception.

Committees of management of Highland institutions often shared a common membership. In Glasgow this was especially true of the Gaelic Chapel Societies, the Glasgow Auxiliary Society, the Glasgow Highland Society and, during the early years of its existence at least, the Gaelic Club of Gentlemen. These Highland societies mirrored those other voluntary societies in nineteenth-century cities that drew together a middle-class élite to exert hegemony over the urban working classes.[74] In the Highland case, the identity of both middle-class élite and those over whom they sought dominance was additionally internally divided on grounds of local Highland or clan loyalties.

Within these formal institutions, prominent individuals (particularly the ministers of Gaelic chapels) operated or supervised systems of support geared to fellow migrants. Lewis Rose testified to Poor Law Commissioners in 1843 how he sought assistance for Highland orphans in Glasgow. The Rev. Norman McLeod gathered donations for the Highland poor.

We have many wealthy Highland merchants in Glasgow...who, though they don't understand Gaelic, attend the church where their fathers worshipped in their own language. They very often give me sums for the Gaelic population. I have many who don't wait to be asked, and they put 5L. into my hand at a time.[75]

McLeod was, additionally, chaplain to the Highland Society's schools, a founding member of the Glasgow Celtic Infirmary (and a director along with Lewis Rose, Lachlan McLean and John Lamont) and of the Highland Stranger's Friend Society, 'a society for aiding poor strangers [Highlanders] who have no legal claim for support upon the public funds of the city'.[76]

McLeod's involvement illustrates the complex way in which migrant culture was made and remade. His role in part depended on active membership of principally middle-class Highland institutions who drew their financial support from wealthy Highland-born to support the migrant poor. He acted as an individual to assist Highlanders: evidence to the 1843 Poor Law Inquiry shows him arranging for proper burials of Highlanders, writing letters to parochial authorities in Glasgow and ministers in the Highlands asking for financial relief for certain individuals, getting Highlanders jobs, arranging entry to Glasgow Royal Infirmary for sick migrants and chastising the nature of the Poor Law system itself.[77] Here was someone, himself Highland-born but demonstrating, too, all those complexities of status and ideology characteristic of the native urban middle classes. The fact of shared birthplace, but not that fact alone, legitimated his voluntary philanthropism towards the Highland poor. And, at the same time, McLeod was minister of a Gaelic chapel where English language services for wealthy Highlanders supported contrasting Highland identity through Gaelic services for the migrant poor. There, and in the Highland Society schools that operated in his chapel and whose chaplain McLeod was, the English language was used for education and cultural change. For McLeod:

Unquestionably, the want of education is one of the greatest evils, and, connected with that, the want of the English Language. The education, as at present pursued, tends to introduce English. We feel the necessity of introducing English, and have taken the way of teaching English through the medium of Gaelic, and that is marvellously successful.[78]

McLeod and others like him may not have had the same degree of legitimating authority within the migrant population that Werly and others have seen the Catholic priest to have for the more tightly knit Irish.[79] They may not have done so precisely because the Highland congregation was scattered and the means to their spiritual supervision and hegemonic dominance less easy than for the spatially concentrated Irish. But it is clear that cultural brokers like McLeod, working within and across migrant institutions, were vital in making Highland identity in Lowland cities.

'Highlandness' depended on more than place of birth. Identifying migrants as Irish, Jewish, Welsh or Highlander by virtue of shared place of birth is a factually based methodological distinction which, as fact, is not open to interpretation. But such distinction may have little actual meaning *within* the group in question, however much it served to distinguish migrants in the eyes of contemporary outsiders and to separate such people for analytical purposes now. Migrant identity has both external and internal meaning. Uncritical attention to a shared ethnicity from which commonly held values are presumed to arise not only focusses on the first meaning to the relative neglect of the second but also ignores the socially constituted processes making up both.

In the case of Highland migrants in nineteenth-century Glasgow, their identity as Celts was in part determined for them prior to their arrival. This external image was additionally falsified by the incorporation into dominant culture of specific 'Highland' attributes: the kilt, whisky, Romanticism and particular character traits. The outsiders' image of the urban Highlander was founded then both on class-related and perceived notions of inferiority and on motivating claims to lack of culture in relation to given 'real' levels of economic and social development.

Highland identity was a class not a mass phenomenon. For lower-class Highlanders, identity was determined by adherence to Gaelic, their relatively subordinate position in the urban labour market, in part by scattered residential distribution and, in part also perhaps, by a conscious opposition from non-Highland-born within the same class. We are told in one contemporary source that 'there are said to be combinations among the operatives [in Glasgow] against them, which they find it difficult to resist'.[80] Other work has denied the existence of persistent anti-Highland sentiment in this way, at least from within the working classes.[81]

Within Glasgow's Highland population the hegemony of the upper classes (Highland-born or not) was realised through processes simultaneously intellectual, spiritual and economic. The first was based upon the motivating need for the lower-class Gaelic Highlander to speak English if he or she were to advance in life (an attribute already gained by many Highlanders wealthy enough to do so). The second was determined by that conjunction of Presbyterianism, philanthropism and self-determining moralism that activated the middle and upper classes in helping the less well-off in nineteenth-century urban Scotland. Economic dominance (and, in turn, the persisting poverty of the majority of the Highland poor) allowed specifically migrant institutions such as the Gaelic chapel to have Gaelic and English services. Wealthy non-Gaelic-speaking Highlanders supported the continued existence of Gaelic-only services. This paradox is doubly important if one admits that the inability of lower-class Highlanders to attend chapel regularly, or even to do so as a whole family, may actually have allowed such persons to remain Gaelic speaking longer.

Language, important as *a* basis to migrant identity for Gaelic Highlanders, could not form *the* basis to a fully articulated Gaelic consciousness or class-based counter-hegemony for those reasons explained above. And even within the population of Gaelic-speaking Highlanders (as within the Highland-born population) there were persons for whom the same attribute did not mean the same thing in terms of migrant identity.

For some migrants a sense of Highland identity through continued use of Gaelic (though not a consciously oppositional stance) was maintained *because* they did just that in contrast to those of their fellows who felt a sense of Highlandness *despite* not speaking Gaelic. Some individuals, notably the ministers, effected a compromise with different interest groups within the migrant population. Men like the Rev. Dr McLeod (known as *Caraid nan Gaidheal* – 'friend of the Highlander' – for his charitable work) looked to the lower-class Highlanders in the ways he did because they were lower-class not alone because they were Highland-born: shared facts of birth were seen not as the basis for a persistent and monolithic 'migrant culture' but only as a means to the hegemonic making and remaking of Highland migrant identity along lines of class.

The foregoing analysis has suggested that attention to ideological relationships and hegemonic structures within migrant populations may be more illuminative of what is meant by migrant identity in the nineteenth-century city than attention to a supposed unity based on the predetermined fact of ethnicity. Processes acting to shape the identity of one group of urban migrants were not the result here of 'the otherness' of Highlanders from hosts but were, in crucial ways, influenced by class-based relations and other distinctions within the migrant groups. These processes were in addition to, not dependent upon, the externally motivated claims of hegemonic blocs in urban society as a whole. For some contemporaries anyway, urban Highlanders were brought to a state of heightened culture by their very presence in cities: 'The Highlanders...became modified by contact with those influences which were lacking in the solitude of their mountain or island homes.'[82] But urbanisation was not this simple: Highland migrant culture made itself as much as it was made.

Acknowledgements
I am deeply indebted to my fellow contributors for their perceptive comments on an earlier draft of this essay and to Bob Morris, Richard Dennis, Geoff Ely and Richard McArthur for their remarks on sections of it. To those archivists who searched (unfortunately without success) for the original 'Gaelic census' returns I owe much, and for permission to quote from sources in their care I acknowledge Strathclyde Regional Archives and the Registrar General for Scotland. Funding for the research on which this paper is based came via a college research grant and I am grateful to my research assistant, Alexandra Watson, for the material used as the basis for Tables 3.2 and 3.3.

4

The country and the city: sexuality and social class in Victorian Scotland

J. A. D. BLAIKIE

Paradoxes pervade our understanding of class: property relations become sacrosanct at a time of violently shifting patterns of ownership; charity co-exists with the harsh exploitation of labour; ideals of benign stability act to disguise bitter transformations. In *The country and the city* Raymond Williams explores the 'structures of retrospect' that have preoccupied socially mobile intermediate groups, those concerned with 'achieving a place in the altering social structure... but continually threatened with losing it'.[1] The moral posturing to which such a temporary foothold has given rise is perhaps most vividly portrayed in a fictitious dichotomy between town and countryside whereby the image of a lost rural innocence serves as an ideal aspiration against which the consequences of rapid and continuing urban growth are measured and found wanting.

In assessing the thinking, and indeed the practices, of an increasingly dominant urban bourgeoisie, we must pay some regard to the ways in which that class sought to consolidate and maintain its separate identity or social distance. A number of studies have discussed this issue, Maclaren and Gray drawing their conclusions from specifically Scottish research.[2] Meanwhile, although class-based notions about housing and public health, for example, are well understood, sexual attitudes have – with occasional exceptions – been dismissed as self-evident abstractions.[3] This is an important oversight, for nowhere more than in the sphere of sexual morality was the contradictory nature of the relationship between classes in Victorian Scotland so clearly evident and nowhere is the relationship more clearly contradictory than in the juxtaposition of urban myths and rural realities.

The following discussion reconstructs a history of social concern over what contemporary moralists regarded as an emergent social problem. The essay examines the socially opaque character of working-class sexual behaviour – its evident resistance to objective contemporary appraisal – and shows that while there was much rhetoric, and some action, no practical changes were effected. As with their counterparts in the turbulent Paris of the same

80

period, the bourgeoisie seem to have been, in Harvey's words, 'fecund with ideas though short on their application'.[4] It would be foolish, then, to accept the 'facts, figures and opinions' cited here as representative of the 'commonsense' of the people. As such, they do not provide particularly accurate information, still less objective accounts of rural working-class attitudes or behaviour. Rather, they represent the conventional wisdom of one class of the population that was particularly vocal in its moral condemnation. The history of sexuality itself cannot be read off from a series of statements of disgust, but we can learn much about the contours of class consciousness from the tracts of the times.

By the mid-nineteenth century the revelation and recognition that, for the first time, the urban population of England was greater than the rural, prompted a series of social warnings that were part and parcel of a novel imagery of the city. Similarly, whereas in 1755 less than a tenth of the Scottish population lived in towns with over 10,000 inhabitants, by 1851 a third, and by 1891 a half did so.[5] Such an unprecedented expansion was greeted with alarm, as in England where:

If middle-class moralities invoked peculiar anxieties, the development of a huge working class throughout the nineteenth century posed immense moral problems of its own. The fundamental problem as conceived by middle-class moralists was the effect of industrialisation and urbanisation, and in particular factory work, on the working-class family and the role of the woman within it.[6]

However, as Williams has noted, 'to understand this whole process more closely we have to push beyond the general classification of "urbanization"'.[7] Statistical concern over living conditions in British towns during the 1830s and 1840s was informed by an ideological commitment to the solution of social problems, its roots lying with a social pathology derived from medicine rather than religion. Although 'definitions of problems and the elaboration of policies for tackling them were sterile exercises in the absence of effective administrative machinery through which remedial action could flow', legislative action was hesitant.[8] Changes came, nevertheless: in Scotland the Poor Law Amendment Act of 1845 and the advent of civil registration of births, marriages and deaths ten years later testified to the slow percolation of new ways of monitoring and categorising the population. Social investigation was a gospel unto itself, but the shocked response of bourgeois Scotland to the fact that in 1855 the national illegitimacy ratio was 7.8 per cent while England's was only 6.4 per cent reflected the combination of a social scientific language of causation with an evangelical morality. Prior to this 'every Scotchman [had been] convinced that his own country was the most moral on the face of the globe'.[9] Now, as the Rev. Dr James Begg retorted:

We could even construct a kind of map exhibiting these results, in a visible form, like the operations of a rain-gauge, upon which even the most obtuse of men could trace connections, which, however little understood, are as well established apart from the omnipotence of Divine grace as the revolutions of the planetary system.[10]

In the later twentieth century unmarried motherhood has gained a good deal of notoriety in government circles as a prime indicator of alienation and family breakdown in an *urban* underclass. While commentators such as Charles Murray have stressed the significance of the city in providing specific conditions through which transmitted deprivation flourishes, others regard the social construction of 'welfare motherhood' as part of an ideological onslaught that has more to do with the generation of social disdain for national policy reasons. Similarly, in the nineteenth century attempts to shift responsibility for the evils of the industrial revolution away from the factory and on to the 'accidentally' related facts of urbanisation had boosted the promotion of an individualistic view of social process that 'blamed the victims' of change rather than its instigators.[11] The environmental origins of moral failure were nonetheless recognized (see Chapter 1). Indeed, as far back as the early 1820s, the famous evangelist and reformer Thomas Chalmers had associated 'improvident marriages' and 'illicit unions' amongst the working classes with the detrimental effects of industry and urban growth. Thus, in his efforts to end pauperism and vice in Glasgow, he continually craved the intimacy and mutual aid of small rural communities.[12] However, the veracity of such comparison was ineluctably undermined by the publication of the Registrar General for Scotland's first detailed illegitimacy returns in 1858. Before that date the image of a pious peasantry had merged with that of the romantic noble savage, making it possible to maintain the fiction that 'outside the worst areas of Edinburgh, Glasgow, or Dundee the cotter was still the real Scot'.[13] Thereafter, the marked concentration of illegitimacy in the agricultural zones of the northeast and southwest, which was clearly portrayed in the statistics, shattered the illusion. The portrait of a rural Arcadia was seen to be a fake: bastardy levels were far higher in the Lowland farming counties than they were in the towns and cities of the central belt. The ways in which middle-class Scotland dealt with this thorn in the side of a philosophy that could otherwise accommodate an individualist notion of commercial progress alongside a conservative and nostalgic social morality illustrate some of the complexities to which Williams has alluded.

'The map of vice in Scotland'[14]

Prostitution, by virtue of its proximity within the cities, represented a specific challenge to the development of middle-class moral supremacy that was far too close for comfort. Ironically, however, bastardy only began to pose a similar threat once it was recognised that illegitimacy was not an urban but a largely rural experience. Although its potentially invasive capacity was evidenced in the steady and increasingly forceful stream of young migrants flowing off the land and into the cities, the provocation for moral outrage was more strictly ideological than territorial. Until now the countryside had been praised as the domain of all that was wholesome; the city, by contrast, had been castigated as a modern Babylon. A comparison of quinquennial bastardy ratios by counties between 1855 and 1939 illustrates a pronounced regional emphasis, with levels consistently highest in the northeastern and southwestern peninsular provinces. In the former, indeed, Banffshire maintained its position as the county with the highest index of extra-marital fertility (I_h) right through from 1861 to 1921.[15] These geographical peculiarities did not fail to strike contemporaries, who were immediately puzzled, if not shocked, to discover that bastardy was a predominantly rural phenomenon associated with the heartland of peasant virtue rather than the supposed hotbeds of urban vice. Explanations abounded, varying from illiteracy, intemperance, hiring markets, imbalanced sex ratios and prostitution to race, religion and – somewhat desperately – rainfall. All could be and were refuted by the Registrar General, who looked then to the nature of accommodation provided in the countryside.[16]

He was not alone. By mid-century the churches identified solidly with the new middle class in the cities in their desire to implant the household mores of the Edinbourgeoisie. Deputations to the countryside inveighed against the bothy (barrack-like accommodation for farmhands) and other systems of farm management that separated servants from their families by failing to provide cottages, whilst investigators became so obsessed with comparisons between farm servants and the relatively virtuous fisherfolk that at times it appeared as if these were 'the only classes with sexual functions'.[17] This focus of concern with rural labour reflected a felt loss of social honour, for now that so many urban workers were outside the church, this was the only sector of the working class over which they might still claim influence. As Muirhead notes, in relation to the sister 'social evil' of prostitution: 'So much which could on a small scale be contained within existing sanctions, grew into fearsome spectres with new economic conditions and population patterns.'[18] The psychological threat posed by the arrival of a displaced pauper host from the rural hinterland was considerable. Begg noted: 'We are in the process of a social revolution which threatens to swallow up the middle-classes...the swarms thus driven into the cities eat up the shopkeepers.' Hence his plea that 'all classes must be made to feel that their paramount interest consists in the building of a strong middle-class...the cheap and strong defence of nations'. With the advent of reliable registration, this imagery of invasion extended to 'immorality':

Now at this rate of 10,000 bastards a year, we shall in ten years have nearly 100,000 bastards, or a population of them equal to that of one of our largest cities; a result sufficiently startling, and well-fitted to alarm all classes of moral and social reformers.[19]

Such were the 'tabular certainties of a new agenda for social reconstruction'.[20] While Chartism, French socialism and other fashionable fears
conspired to inhabit the chamber of horrors, statistics played a pivotal role
in locating pockets of contagion. The northeast, having been targeted by
Seton as a 'black spot' on the map, was found to contain adjacent parishes
with combined bastardy ratios exceeding 23 per cent. Here no parish had
returned a ratio of less than five per cent and eighteen parishes recorded over
20 per cent. Most scandalous of all were inland farming parishes (rising to
over 40 per cent); fishing villages contributed contrastingly low quotas.[21]
Whereas in certain districts 'the very air seem[ed] impregnated with the
germs of immorality', the visitor to Banffshire was 'sometimes startled as his
host points out to him 15 farm houses along a countryside ... each of which
shows, or will show within a few months, the track of the destroyer'.[22]
'Scotia is sick', said the Registration Examiner in 1861. 'Already have her
physicians, diplomatised and undiplomatised, written off no end of
prescriptions'.[23]

Discussion of rural sexual behaviour was increasingly couched in terms of
farmworkers *as a class*, strengthening thereby the middle-class assumption
that they were a 'race' apart. Conversely, the middle-class desire for
hegemony demanded that such an anomalous distinction be rectified. Here,
housing and courtship provided the active loci for reform.

Amply codified in the annual reports of the Religion and Morals
Committee of the Free Church and its Housing Committee, the Church of
Scotland Life and Work Committee and its Commission on the Religious
Condition of the People, Pastoral Letters on the Increase of Immorality in
Rural Districts, Synod Reports and Addresses, a plethora of private
broadsides and the Transactions of the Highland Society, among others, the
orthodox litany of moral concern over 'the family of sins against the sanctity
of sex' guaranteed that the 'illicit' activities of farm servants were to reach
the privileged status of a social problem. There were two reasons behind this
unprecedented scale of investigative work. Firstly, as we have noted, the
regional distribution of bastardy levels served to undermine the myth of rural
virtue and, in so doing, ruptured the complacency of an ecclesiastical vision
that linked sin with the twin spectres of industry and urbanisation. Secondly,
the pronounced occupational concentration of the 'vice' amongst farm
servants and the remarkable divergence from middle-class courtship patterns
forced the realisation that the sexual behaviour of this rural workforce did
not conform to the respectable stereotype. These factors ensured that the
imagery of illegitimacy was that of invasion and that the ideology offered in
response was one of containment. If the parental hearth, 'divinely
appointed', was the bulwark of stability and social order, then any pre- or
extra-marital departure represented an index of potential chaos.

Contexts for control

If the development of capitalism transformed the cities by creating a vast, if differentiated, urban proletariat, it also reshaped rural society. As Gray, Carter and others have demonstrated, the century or so after 1775 was a period of drastic change in a previously recalcitrant agrarian society.[24] In northeast Scotland landlord demolition of cottar houses after the passing of the Poor Law Act (1845) had filled the towns with itinerant paupers, while rural districts became 'more degraded and debased than the cities themselves'.[25] The Free Presbytery of Strathbogie complained, in 1858, that accommodation shortages forced men in farm service to live apart from their families who inhabited nearby villages, which were simultaneously foci for single mothers weaning their bastards in low-rent lodgings.[26] Their fortunes and frustrations reverberated throughout society, one tremor being felt in the endemic rise in bastardy across much of the region as the opportunities for independent family formation regressed.

Kussmaul has shown living-in farm service to be a youthful, transitional stage in early modern England.[27] In northeast Scotland in the nineteenth century the life-cycle stage between fifteen and twenty-four was similarly characterised by living-in service and high levels of localised mobility. However, the shortage of leases meant that increasingly, rather than settling back into peasant stability and thus rekindling the family cycle, members of this cohort were forced to adopt new strategies: some left the land entirely whilst others continued awhile in service hoping that, eventually, crofts would become available. Such aspirations became increasingly attenuated as fewer and fewer steadings came on the market. It has been argued that life-cycle servants out with the parental eye courted freely in the anticipation that pregnancy would precede marriage. However, when agrarian crisis came to restrict opportunities for couples to set up independent holdings, the margins of insecurity involved in risking intercourse widened dramatically to embrace not only the very poor and those dynasties through which bastardy followed a line of descent but also considerable numbers of those who were now potentially propertyless.[28] Illegitimacy is not just about sexual relations; it concerns the whole process of production and reproduction. However prescriptive or pervasive cultural norms may have been, they were defined, negotiated and redefined according to specific material circumstances. The need to understand the complex realities of such a process cannot be emphasised too strongly. Moralists nonetheless continued to rely on repeated appeals to a generalised and illusory past. Whilst popular novels and journalism eulogised an unchanging moral community, Rural Police Committees, kirk sessions and philanthropists sought to contain the 'migratory habits' of farm and domestic servants who managed to evade surveillance by parents and employers.

Initially centring about the arguments between Begg – via the Free Church Housing Committee – and the Registrar General, the accommodation of unmarried men became the object of most controversy.[29] From the vantage point of metropolitan Edinburgh, bothies were seen as a root cause of the problem, founded as they were on the 'subversion of the principle of the family' where Providence had included servants in order to establish 'kindly relations' between masters and men.[30] Not only this, but the lack of cottage accommodation for those wishing to marry exacerbated the situation; hence the claims of a Banffshire general practitioner that men were sometimes placed in the circumstance of having a double family to provide for, 'the earlier portion being illegitimate', since 'when the first offence occurred, they had not the means of making the proper amends by marrying and sheltering mother and child in a dwelling of their own'.[31] Similarly, Begg's crusade was not directed at farm servants *per se* but against the prevailing agricultural system, a system which, had it been predicated upon the soil being the right of the people and not the property of rapacious landlords, would have served to resurrect, he claimed, the happy familiarity of peasant democracy. As it was, the young ploughman could no longer aspire to the simple prospect of 'a good wife, a good cow, and a good razor'. Rather, it was claimed, 'if circumstances over which he has no control, hinder his marrying in his lusty prime, he is tolerably certain to do something worse – to show, in fact, his contempt for Malthus in an illegitimate fashion'.[32]

The bothy had been an object for anxiety in philanthropic circles since the early 1840s when the enumeration of types of farm servant accommodation had produced corresponding lists of corruption and depravity.[33] But with Begg's supposition that 'bothies cause bastards' ably and authoritatively defeated on the Registrar General's evidence, the discovery that there were 2,170 bothies in the Angus and Mearns Synod boundaries or 600 women housed in Lothian bothies no longer provided sound ammunition: neither district possessed the notoriety of Aberdeen or Banffshire – counties where the bothy was relatively scarce.[34]

The debate continued, with emphasis shifting 'from the buildings themselves to the whole problem of the social context of the worker living away from home under circumstances where there was no alternative parental control from the farmer'.[35] In taking for granted the improbability of self-restraint, however, this line of reasoning betrayed an evasion: possible alternative motives behind sexual conduct were not countenanced. Commentators were rarely content to argue wholly from the standpoint of environmentalism; and in some cases where individual moral failure was not explicitly invoked hypotheses verged on the ludicrous:

I think that the excessive illegitimacy in this county [Aberdeenshire] is due partly from the men and women being associated together both in the kitchen and also in the fields, and also to a great deal of oatmeal being consumed here, which contains a great deal of phosphorus and has a tendency to inflame the passions.[36]

Things were just not that simple and, as Boyd remarks, if the map of 'connections and causes' that Begg wanted could have been produced, 'he would doubtless have been infuriated by the complexity of roads, lanes and sheep tracks leading to illegitimacy'.[37]

More tantalising evidence came from the revelations that rural working-class wooing was decidedly different in character from the respectable stereotype. The transfer of property depended on a woman's chastity, but amongst the propertyless bridal pregnancy was the norm. An important intervention in the contemporary discussion of 'immorality' came when Dr Strachan of Dollar published the findings of his investigation into premarital conception. Extrapolating from a study of thirteen parishes, Strachan, a country doctor with a large midwifery practice, had concluded that an estimated 61,000 women in Scotland had been unchaste in one year, 17,000 bearing bastards, the remainder marrying prior to childbirth. His major finding, however, was not the high bridal pregnancy itself but the 'astounding' discovery that this pattern was almost wholly confined to the rural working class. In the rural parishes he studied, over 98 per cent of bastards and children born within the first six months of marriage had working-class parents. On the basis of these figures he attributed extramarital fertility to class differences in courtship behaviour: whereas middle-class courtship was 'open and honourable', for farm servants it was clandestine, consisting of 'stolen interviews', usually at night. Ironically, this was necessitated by the 'no followers' rule imposed by mistresses and the absence of liberty for 'acquaintances and sweethearts to meet in the family circle'. Since young men and women were not allowed to meet in private, social control itself drove sexual relations underground. Strachan proclaimed that since 'the great cause of this immorality is coarseness and indelicacy of manners it should prevail less amongst skilled artisans...most amongst agricultural labourers'.[38] Yet, responded an Aberdeen journalist:

With them the vice is gross and open; we see it in its full extent and know its full limits...It is not amongst them that you find one class treated with the honour and respect due to superior beings, and another class deemed fit only to minister to the vilest passions, and then with the tacit approval of a 'society' composed to one half of their socially superior sisters.[39]

Since family fortunes hinged upon the prudent choice of partners, middle-class courtship demanded an open etiquette, whilst indiscretions were concealed, taboo; with the rural working class, where no such constraints operated, the situation was reversed. The Victorian double standard, which demanded that daughters of the middle class remain virginal until marriage whilst young men were expected to have gained sexual experience, ensured that prostitutes would always find a market among the sons of the bourgeoisie.[40] However, whereas the early Victorian investigator Tait (1842) claimed that 80 per cent of the seducers of unmarried women in Edinburgh had been relatively wealthy men, the later Victorian pamphleteer and surgeon, Vacher (1867), studied the occupations of seducers of 364 unmarried women who had delivered their first child in the Royal Maternity Hospital and found that 'a very trifling per cent of the seduced have been led astray by men moving in a higher sphere than themselves [but] as a rule the seducers in each grade of the community are to be found within that grade'.[41] Either thinking had shifted away from the acceptance of hypocrisy towards an attempt to redirect the accusation towards the so-called 'respectable' working class, or, within a generation, the pattern really had changed. Not only did the towns contain more members of the middle classes, suggested Strachan, but 'a large proportion of the working men consists of skilled artisans, whose position and whose feelings and manners approximate to the class above them'.[42]

While such imitative behaviour might be seen as a 'negotiated redefinition' of middle-class conventions,[43] the protocols forming the customary bedrock of rural society seem to have predated nineteenth-century class differences. Both church and state had opposed this divergence for centuries and, to this extent, it could not be regarded as a recent development, an 'index of "disorganisation" in an urbanising epoch.'[44] The fresh element in the equation was, again, the introduction of official statistics in the 1850s. The markedly regional and overwhelmingly rural distribution of high illegitimacy weakened the impact of a calculus that had increasingly blamed individual transgression of social norms rather than the social system for Scotland's moral ills. The implication was that this courtship pattern reflected a '"native" rural custom, upheld within an integrated social and cultural system of norms' distinct from those of the dominant culture. Marshalling a superb welter of statistical material, the Banffshire schoolmaster William Cramond remarked that almost one bastard per day was being born in the county and that in less than a generation as many illegitimates had been born in the county as the entire combined population of the towns of Banff and Keith. Alongside the now familiar speculative catalogue of causes, the 'low moral tone prevalent in the County' remained as high on the list of speculative causes sent in by the respondents to Cramond's questionnaire as it had been twenty-four years previously when Charles conducted a similar, though far less exhaustive survey of local opinion in the southwest.[45] Though scarcely any female servants were asked their opinion, Cramond's interviewees supplied important leads, such as:

I have remonstrated with mothers about their daughters keeping bad hours and bad company, but their general idea seemed to be that girls must take their chance...

I have heard many a mother of this kind say: 'It's nae sae bad as stealing or deein' awa' with the puir craters'.

I do not think that in one case out of every six that I have dealt with, has ever marriage been referred to or looked for by the woman.[46]

Such commentary spoke not of individual transgression shamefully repented but of culturally condoned social acts. Indeed, reconstitution analysis of northeast Scottish communities suggests that endemic bridal pregnancy may well have reflected an age-old system of 'fertility testing' among the peasantry, but that when, as in the latter half of the nineteenth century, an intending couple no longer had the prospect of a home in which to raise a family, illegitimates would ensue.[47]

By comparison with the Lothians or the English Corn Belt, the incursion of capitalism here had not occasioned a sharply differentiated arrangement opposing wealthy farmers to a mass of landless labourers. Instead, the intermediate strata remained, many holding land but also providing labour through their sons and daughters for those above them. The slow drift towards polarisation gained momentum after 1870, when, during the 'Great Depression' the enforced superexploitation of family labour resulted in the outmigration of children in search of better conditions, whilst the importation of Irish and Canadian stores undercut the indispensability of cattle rearing on the smaller farms.[48] As engrossment penetrated more deeply, shortages of leasehold accommodation began to disrupt a pattern of reabsorption whereby all 'roared around the "ancient strange whirlimagig" of the old lowland Scots peasantry from labour on the paternal holding to hired farm service and then, for a proportion of any cohort of peasant children, to day labouring and back to peasant farming'.[49] On reaching marriageable age, more and more sons and daughters left for alternative occupations if not for another area. Others were left literally holding the baby.

Contagion and containment

In *AIDS and its metaphors* Susan Sontag discusses the characteristic historical imagery of sexually transmitted diseases. The symbolic disdain for such illnesses has been one that reflects the threat posed to dominant groups in society by the 'deviant' behaviours of minority groups failing to collude with social norms of appropriate family formation and maintenance. AIDS, like syphilis before it, is surrounded with the nomenclature of 'plague', of judgements on a community of outsiders for 'evil' doings, of a roster of tell-tale symptoms of contamination, characterised temporally in terms of progressive stages of decay, and spatially as foreign, invasive, and in need of surveillance and containment by quasi-militaristic strategies that act to forestall the virus by stimulating the immune system – social as well as medical – in advance of attack.[50] As will already be apparent, such a 'metaphor' has clear resonances when applied to the imagery of sexual conduct in nineteenth-century Scotland. This is hardly surprising, given that venereal disease certainly provoked fears of contagion via middle-class sons who engaged prostitutes. As Smout notes: 'This anxiety might have been based on a real increase in the danger to public health: prostitutes who formed the pool of infection were almost exclusively urban, and as urbanisation developed, prostitution and VD could be expected to grow in proportion.[51] For our purposes, what is of particular interest is the dialectic posed between urban and rural spheres. Although scarcely confronted directly, it is clear that migration from the country to the city forms one dynamic that consistently foils attempts at moral containment.

As the cities expanded attempts were made to create a substitute home environment via industrial schools, reformatories and emigration schemes that rounded up ragged children and relocated them with 'respectable' farmers in Canada.[52] Meanwhile, prostitution was rife. 'Rescue' workers saw domestic service as a solution, yet this was in fact one of the main sources of recruitment: 81 per cent of applicants to the Edinburgh Magdalene Institution between 1855 and 1860 were one-time domestics. 'Immorality' was therefore 'rampant in that class upon which much of the happiness of the community depends and whose interests are so intimately bound up with our own', whilst anxieties arose because 'our maids fill[ed] the country with illegitimate children and swarm[ed] our streets with prostitutes'.[53] Given the contradictions of the Victorian double standard, however, it was perhaps a greater evil to bring the facts of prostitution to light than to ignore its existence. Since systematic monitoring was also extremely difficult, there was no Chadwickian movement to investigate vice in all its material complexity. Still less attention was paid to the relatively low bastardy levels in the towns, despite a consensus that a worse sexual morality prevailed there than in the country. Instead, the vocabulary of 'immorality' was one suffused with epidemiological analogy. Though not medically dangerous *per se*, illegitimacy and prostitution had become metaphors for social disease in nineteenth-century Scotland. Invoking parasitism, one reformer thus remarked that:

We look with loathing upon a patient covered by a disgusting malady; but that is never admitted as a reason for refusing to apply the necessary remedies, nor ought to be in this disgusting moral disease... These degraded women must each of them be considered as a centre of evil, injuring, or it may be destroying, many around her... a woman who is on the streets lives and must live by the corruption of others. No other mode of livelihood is open to her. If she cannot lead men astray, she starves.[54]

Given the ideological potency of such metaphors, it is instructive to compare 'immorality' with other epidemics. According to Allan Maclaren, analysing nineteenth-century Aberdeen: 'It was precisely because cholera was community-located rather than located in a specific class that class perceptions were intensified.'[55] While both prostitution and illegitimacy were very clearly class-specific, it was no simpler to throw cordons sanitaires around controllable 'pockets of contagion'. If, as was demonstrated, prostitution was absent from the countryside but endemic in the cities and illegitimacy rife in the hinterland but low in the towns, how could the two phenomena be connected as part of a unitary framework for degeneracy? From the 1840s it had been recognised that the demand for female domestic servants and women workers in the factories had fuelled much of the immigration to larger urban centres. Two lines of reasoning followed. First, the subterranean taints of overcrowding and bad company coalesced in the shape of the common lodging houses, 'the media through which the newly arrived immigrants find their way to the Fever Hospital'.[56] By this argument, reflected in such broadsides as Shadow's *Midnight scenes and social photographs* (1858), prostitution and intemperance in 'poverty's lowest valley' became the lot of the unfortunate and once innocent rural maid corrupted by the city. The second argument, antithetical to this, lay blame squarely at the feet of the rural migrant. In *The state of St David's parish* (1841), Lewis remarked that:

It is not the native population but the imports of Dundee during the last twenty years that have changed its character. Its rising trade and manufacturing enterprise attracted for many years all comers, with a character or without one...When they came and settled the Church and country were asleep and unprepared to bless them.[57]

Later, 'scientific' arguments suggesting demographic determinism were propounded in an attempt to explain away the difficulties faced by the churches:

With increasing density of population, illegitimacy also increases, and moral pressure becomes uncertain in its operation and difficult of application. In Banffshire, therefore, where the density of population cannot now be less than eighty to the square mile, the sum of illegitimacy is...very great; but when this density arrives at its maximum, as in the case of a large town, illegitimacy, which is the result of one act, or occasional acts of incontinence, is superseded, although not entirely, by prostitution, or habitual incontinence.[58]

Since lower urban illegitimacy could not, on these accounts, be attributed to a sterner moral code, it might have been caused by higher perinatal mortality, with all its connotations of child murder, or by poor registration coverage allowing concealment. One commentator, an advocate, argued, somewhat coyly, that:

There are unquestionably worse forms of vice than illegitimacy and it is well known that one of the most deplorable of these...is invariably attended by no tangible results...If illegitimacy were to be taken as the gauge of morals, it is to be feared that very false inferences would in many instances be drawn; and, guided by the statistics alone, the most casual observer would hesitate to conclude that, in purity of manners, Palermo, Hamburg and Amsterdam are fully three, and London, eight times better than Paris, Berlin and Vienna.[59]

Although observers were not 'guided by the statistics alone', it was the impossibility of quantitatively assessing these 'unquestionably worse' vices that often led them to impute their arguments from the known facts of illegitimacy. As a result, illegitimacy was used as a moral gauge. For example, the Sheriff-Substitute for Aberdeenshire contended that although statistics for infanticide were 'unobtainable', 'in the case of lawful children it can hardly be said to exist at all now. It is in the case only of illegitimate children that it still exists, but it exists there to a serious extent... the mothers and fathers of illegitimate children form the lowest class of society which is higher than the habitual criminal population.'[60] Sporadic evidence that relatively high numbers of bastards died in infancy had been gleaned. In one registration district of Dundee in 1855, thirty-two children were reported to have died at birth and nineteen from within five minutes to one hour after. Medical attendants had been present in only five instances, the remaining forty-six being attended by neighbours or midwives. In his notes the District Examiner referred to the term 'bowelhive' being used as a convenient term for child smothering, adding that: 'The informants in these cases are commonly elderly women of suspicious appearance and character, who had been present at the time of birth, and who can scarce tell their errand to the Registrar without betraying a guilty blush.'[61]

Such reasoning signifies important elements in the model of 'immorality' being constructed. Cities were evil places. Since, by reference to birth statistics alone, they exhibited relatively low bastardy levels, it was assumed that within the impenetrable dens of vice underclass mothers were resorting to child murder. In the absence of corroborated 'expert' opinion, the imputed bad character of the lower orders had to be invoked. Poverty, poor hygiene and attendant malnutrition were later countenanced as causes, but only within the context of a prevailing 'low moral tone'. What remained unresolved was the genesis of such an attitude. Whilst an ideal moral consensus may have been sought in the shoring-up of 'respectable' attitudes towards sex, marriage and the family, this model perched uneasily atop a mass of facts, figures and opinions piling up in Edinburgh as the fruits of investigations and explorations found their way to the General Assemblies of the various churches.

Philanthropy, urban and rural

Official interest reflected a perceived need to contain sexuality, where possible by institutional means. In the later 1850s, with an illegitimacy ratio of 15.2 per cent, compared to Glasgow's 7.5 per cent, Edinburgh's 8.2 per cent and Dundee's 10.4 per cent, Aberdeen was said to occupy the 'pedestal of infamy' regarding immorality.[62] In 1852 Alexander Thomson, convener of the Prison Board, lamented the failure of city missions, Sabbath schools, emigration schemes and prisons to remedy the ills attendant upon juvenile vagrancy and crime, praising meantime the Aberdeen Experiment, by which he referred to Sheriff Watson's pioneering industrial schools.[63] The same year he drafted a Rural Police Committee Report condemning *inter alia* the mobility and immorality encouraged by feeing (hiring) markets and proposing in lieu a registration system together with long-service prizes.[64] He later produced a paper on the Repression of Prostitution, which noted that girls left reform schools whilst still in their mid-teens and were thus prone to temptation, a folly that meant that 'illegitimate children furnish[ed] more than their numerical proportion of the inmates of our prisons', daughters being 'apt to follow the evil example of their mothers'. Discussing 'these unfortunate illegitimate children ... one of the great suppliers of criminals in later life', he exploited this connection as part of an incipient 'transmitted deprivation' thesis.[65]

A less conjectural interpretation was offered by Watson, who, in a detailed study of the city parishes of St Nicholas and Old Machar, suggested that 'vice' was primarily the result of trade crises, factory closures and the consequently distressed position of female operatives:

The high price of provisions in 1853, 4 and 5, and the shutting up of several factories in 1854, by which upwards of 500 females were again thrown out of employment, rendered the condition of these women truly deplorable. They were said to be able-bodied, and so had no real right to parish relief and many of them would have died of hunger, had they not got a supply of bread and broth daily from the soup kitchen, which was kept open from August 1854 to April 1855, and during that time upwards of 70,000 rations were served out. But this supply was insufficient for their support ... and many of them were driven by dire necessity to prostitution and gave birth to illegitimate children. It was the poorest ... that fell into this stage of degradation, and it was ascertained that of 272 illegitimates born in Aberdeen during 1855, the mothers of 101 of them were factory servants.

The mean bastardy ratio for St Nicholas from 1856-60 was 16.2 per cent, for Old Machar, 13.2 per cent. Watson thus deduced that urban bastardy was due to 'the want of oversight and proper supervision for factory girls when out of employment'.[66] The roots of rural illegitimacy, however, were seen to lie in the want of suitable accommodation for intending couples to settle into. Commenting upon Aberdeen and Banffshire's vagrancy figures from 1861 to 1875, he quoted a Police Inspector's remark that 'the children of a new generation are fast treading in the footsteps of their parents, most of them without education of any kind except begging and pilfering'. Unempowered to alter housing arrangements, Watson advocated two reforms: a return to the voluntary system of outdoor relief for single mothers and a better moral and religious education for the young. Principles of residential social work and missionary endeavour continued to stimulate developments in Aberdeen itself, but it was less easy to divert the forces of state legislation enshrined in the Poor Law Act as it operated in the countryside.[67]

In 1859 Valentine told the British Association that in Aberdeen there had not been a single bastard born in the principal streets of the city, the mothers all residing in the poorer districts. Many of these women had been domestics employed in the fashionable districts of the city but forced to leave their situations when pregnancy had been discovered. Analysis of the recorded residences of the mothers of bastards registered from January to June 1859, said Valentine:

shows very distinctly that in the districts where the lower class of houses are situated, there does illegitimacy most prevail. No case is reported to have occurred in Union St, King St, Crown St, Dee St, Bon-Accord St, Bon-Accord Tce, Albyn Place, Victoria St, Union Tce, or Skene Tce.

All the desirable districts were free from blight. Unlike other epidemics such as cholera, therefore, illegitimacy was contained within a tight perimeter of working-class slums.[68] On the other hand, Thomson asserted that 'the number of these births in towns is considerably increased by unmarried females endeavouring to hide their shame by leaving their homes in the country and seeking shelter in the towns at the time of their confinements'.[69] Conversely, it has been shown that although pregnant girls moved into the cities, failure to obtain settlements necessitated removals back to parishes of origin prior to confinement. Such a process doubtless acted as a brake against the further influx of migrant mothers-to-be.[70] In a depressed countryside where cottage demolition continued apace as engrossing lairds attempted to combat the rise in paupers falling on the rates, there was no easy remedy. Although the fact that relief allowances were granted to unmarried mothers with more than one dependent child might have acted as a spur to repeated licentiousness, Watson also pointed out that: 'Vagrancy will never be put a stop to, or greatly reduced, so long as... mothers of young children depending on them for support are offered the [poor]house, which they would starve rather than enter.'[71] However, rather than emphasise the dynamics of such mobility as parts of a single theory of transition, he concluded that illegitimacy was due to 'poverty and want of employment in the towns, and deficiency of house accommodation in the country'.[72]

During the closing decades of the nineteenth century the impact of agrarian depression began to tell upon all classes. Age-selective outmigration clearly presented the rural bourgeoisie with labour problems. By the 1890s the exodus of young women to domestic service in the towns from the northeast hinterland showed that 'no well-educated girl with a spark of ambition... need toil among the "tatties"'.[73] This factor, alongside the apparent leniency of Poor Law Inspectors in the region and the readiness of farmers to engage unmarried mothers, invited fears of racial deterioration, as young men, also frustrated at the lack of croft holdings, drifted into the army and the police force. As a Banffshire clergyman commented:

Many who are now [1888] in service – indeed the great majority of them – are those who by training, physique, character or education are unable to do anything else – not fit for trades, shops, &c; indeed they are the illegitimates of the country.[74]

The logic of such statements was that those left behind on the land formed a pool of degenerates. Ironically, such social Darwinism ran at once directly counter to the two key bourgeois myths: that of a lost Merrie Scotland and that which pictured the city as a modern Babylon.

The contradictions of such a situation did not go unnoticed by reformers. In 1881 a number of ladies, mainly estate tenants' wives, met together at Haddo House. Their object was not only to retain a potential workforce but also to discuss what could be done to raise the morality and living standards of local farm servant girls. Shortly afterwards the Onward and Upward Association was formed, with Lady Aberdeen, who had herself worked to rescue Strand prostitutes, as its president. Domestic work had to be made to appear both a spiritual and a natural calling:

It needs a great deal to be a good servant... It is a high and dignified thing... We want to get the idea of domestic service into the heads of our girls...we want them to realise that it is a position of great dignity and honour...If we can succeed in getting them to understand this idea of domestic service, we shall find much less unwillingness amongst our girls to go into that profession.[75]

Since farm servant girls did not have free evenings and, it was felt, were 'sure to get into mischief', a tutorial system was devised whereby mistresses on outlying farms were invited to distribute question papers and advise on all subjects from Bible history to embroidery, with supervision by Local Branch Committees. Ostensibly a social exercise in outreach education, such activity had deeper-lying normative and material aims. The 'mischievous craving for independence' among the rising generation now filling factories and workshops in the towns rendered farm service, with its panoply of restrictions, a socially inferior occupation. Young women were leaving the land. With those maidservants that remained being in a seller's market, Lady Aberdeen found that:

Everywhere you hear the same lament, that you must put up with untrained, independent lasses, who will do their work in their own way, who will make their own terms...and who will surely desire to flit when the term comes round.[76]

Attempts to reverse this dilution in the quality of domestic labour sometimes took the form of deliberately exaggerated scaremongering. The *Onward and Upward* magazine, retained as a fireside friend for those 'heart homeless' in service on lonely farms, included a series of carefully contrived parables written for the magazine by local clerics. Here, eschatological aspects assumed a drastic significance. In 'Sketches from my visiting book', by 'a country minister', Mary Broon strays from the straight and narrow

path and runs away to lodgings in Aberdeen. Her father, having found her
in childbed, is about to leave when:

A scream from Mrs Walker [the landlady] of 'Come back! Come back!' caused him
to retrace his steps ... There lay his daughter – with the red blood streaming from her
mouth and nostrils – dying.[77]

Whilst such a story may gain some credence from the fact that an 1890 police
report stated that of Aberdeen's 180 prostitutes, 80 per cent were former
domestic servants and were likely to die within four years of entering the
trade, the inference, explicitly stated in the previous tale, was that: 'Sin when
it is finished bringeth forth death.' In an attempt to convey the direct, but
hidden, relation between the 'Two phases of the social evil', the Registration
Examiner had remarked that 'a lady's eyes are not often troubled with sights
of misery. Yet they would stream tears on beholding the body of a former
domestic become the burden of a dissecting table.'[78]

The desire to implant a domestic ideology produced a twofold emphasis:
on the one hand, *Onward and Upward* dealt with the bond between mistresses
and servants; on the other, it dealt with the relationship between mothers
and their families, producing an encyclopaedic range of 'Household hints',
'Homemaker's corner' and the like. Both ways, it dealt with 'the work of the
home which must always be women's first mission'. Seeking pleasure outside
that domain was considered anathema:

Home sweet home! as we love to call it ... is not what it used to be – a mighty power
in the nation. We must try to keep up its strong influence, however. We dare not let
it go, swept away by the love of excitement and variety, so prevalent in the present
day.[79]

Paradoxically, although the Haddo House Association had set out with the
aim of containing the supply of single servants, it rapidly found its attention
being focussed on those women who were no longer vulnerable girls but
safely espoused young women yoked to other forms of domestic labour. This
swing was largely a reflection of the relative ease of inculcating pride in one's
home among those who had an abode to call their own. Meanwhile, for the
single members the introduction of a sliding scale of long-service medals did
not create the desired fashion for remaining in the same situation. In 1891 the
editor reminded those who were moving at the November term to notify
Branch Secretaries of changes of address, adding that those who went to
districts where no Branch existed should become Isolated Associates. By the
next May term she warned that 'there must be no gap in our ranks ... we must
all try to get new recruits'. Retaining a hold was not easy; as one Branch
Secretary remarked:

Owing to a great many removals in the parish, our numbers are not increased ... about
20 of the unmarried Associates (from 56) have sent in no work this year, and only
two of the defaulters have sent their sixpences. I hope the others will not forget.[80]

But, of course, many did. And, yet again, the 'migratory habits' of youth were held up as the culprit. Discussing the 'fallen' who 'do not seem to know they have sinned', Lady Aberdeen noted: 'We do our best to break down their pride, their carelessness, their determination to amuse themselves and go their own way beholden to no-one.'[81] Yet the shift away from concerns about the perils of late adolescence seemed complete when, in 1896, the new editor undertook to run *Onward and Upward* as a 'mother's magazine'.

Against the seductive image of escape from an unremitting round of arduous toil to the heady pleasures of city life was set the couthy morality of Home-Sweet-Home, with the terrors of maternal mortality vividly invoked as a disincentive to flee this cosy nest. Such were the putative 'push' and 'pull' factors, the constituent 'cross pressures of two ways of apprehending the world'.[82] In an effort to enlist individual guilt feelings the material constraints that underlay outmigration were scrupulously avoided. Intriguingly, while the delinquency of the young male farm servant had been excused in part by the absence of suitable accommodation, the attempt to inculcate domestic pride among teenage women ignored this issue.

'Blaming the victim'

But the stigma, the defect, the fatal difference – though derived in the past from environmental forces – is still located within the victim, inside his skin. With such an elegant formulation, the humanitarian can have it both ways. He can, all at the same time, concentrate his charitable interest on the defects of the victim, condemn the vague social and environmental stresses that produced the defect (some time ago) and ignore the continuing effect of victimising social forces (right now). It is brilliant ideology for justifying a perverse form of social action designed to change, not society, as one might expect, but rather society's victim.[83]

Perhaps it would be anachronistic to regard the state of youth in nineteenth-century Scotland as a barometer of the perceived state of the nation. Nevertheless, to use Stan Cohen's term, the younger generation in Scottish agriculture and domestic service were the 'folk devils' of their time.[84] But whatever 'moral panics' ensued, they were certainly as much a reflection of class prejudice as of generational distance. In a country that danced increasingly to the hazardous rhythms of urban commercial expansion, an ascendant but still unsteady bourgeoisie attempted to distance itself from the noisy social problems created by such growth. The publication of the Registrar General's first detailed regional bastardy statistics in 1858 acted as the springboard for a variety of philanthropic proposals, none of which translated itself into effective practice. The class nature of such efforts, it has been argued, is crucially important in understanding why each strategy failed. Victorian moralists did not so much seek the causes of 'immorality' as displace the blame on to identifiable social groups. In so doing they inevitably manufactured a cultural resistance that transcended all reality. To fuse rural bastardy and urban prostitution in the vision of a twinned 'social evil' was to encounter the contradictions of capital: the sexual expectations of the middle-class male sustained prostitution whilst the hiring practices of capitalist farming exacerbated illegitimacy. Meanwhile, appeals to the moral sensibilities of the lower orders were unlikely to succeed so long as the material choices open to them were ignored. In an area where no homes were to be had for intending couples, attempts by the urban-based bourgeoisie to graft their own domestic model on to the social relations of service in the countryside were doomed to failure. They acted to suppress the symptoms rather than the cause by displacing the blame for the evils of capitalist penetration on to the imputed moral shortcomings of its victims. As a result, the written history of concern tells us more about the presuppositions and ideology of its proponents than it does about the conduct of their subjects.

Clerical interest in part followed from a growing unease over the accommodation of farmworkers, although the various church committees, having emphasised the links between housing and morality, went on to develop an aetiology of bastardy far more comprehensive in scope. A specific set of moral and material causes combined to frame farm servants as the social group consistently manifesting an array of pathological symptoms. This particular Aunt Sally served as a counter-image against which an ideology of the 'traditional' family could be vigorously constructed. The particular geography of 'immorality' ensured that, within the northeast, administrators, clerics, doctors, journalists and schoolmasters formed a community of interest which, despite the inauguration of active agencies to combat the 'social evil', brought little pressure to bear.

As McGregor has noted, a prerequisite of social policy was the willingness and ability to define social problems. It may have been the case that, try as they might, well-intentioned moralists just could not unearth the one true cause of the 'Great sin of Banffshire'.[85] Although this may appear to be a plausible explanation, it has been argued here that these commentators, far from representing a perplexed or nonchalant audience, succeeded in intricately constructing a detailed social model of illegitimacy. The failure of all strategies to inoculate rural society against this moral disease, as they saw it, lay not so much in their inability to comprehend the 'problem' as in the material impossibility of grafting a middle-class family ideal on to the social relations of service in the latter half of the nineteenth century. To this extent, the ideological division that middle-class Scotland drew between private vices and public virtue, and which was exemplified in a code of conduct where grave social consequences flowed from individual moral failings, must be examined not only in terms of its own internal logic but also against alternative protocols.

Not least significant here were the differences between urban and rural fractions of the dominant class as regards church discipline for 'fornication'.

Certainly, in parts of rural Banffshire, the pattern of office-holding among lay elders in the kirk broke distinctly with the biographical characteristics of the urban vestries of Aberdeen, where a successful business career was a prerequisite of affiliation. Twelve of the twenty-four elders ordained in Rothiemay between 1832 and 1894 were connected via kin links or themselves directly contributed to the 'immorality' of the parish, nine being the grandfathers of bastards. This picture does not rest comfortably with the received image of the Scottish kirk elder as exemplary, pious and mean. The social class composition of these men was particularly homogenous: eighteen elders were farmers and the rest artisans. Unlike the resolutely middle-class characters of the urban vestries, however, most seem to have been small or middling tenants rather than large-scale rural capitalists.[86] We are not witnessing here a realignment along class lines whereby a new model bourgeoisie adopts the pious church code whilst the proletarian masses lapse into heathenism. In the agrarian hinterland the old moral economy held out against the political economy of class separation. Rising illegitimacy was perhaps one indicator of its incipient decay, but it was not a predicament with which the community's moral custodians were completely out of sympathy.

It was in the interests of both civic order and public health to clean up the streets by investigative action and publicity, prevention, cure and, failing all these, tighter policing. However, as Checkland has pointed out, the official attempt to pursue the prostitute was 'uniform in its ineffectiveness'. Since the problem was conceived 'as an individual moral failing from which social consequences flowed', no clear administrative machinery could evolve through which reform could be managed.[87] Moreover, urban prostitution, insofar as it neatly exemplified the workings of the double standard across the class and gender divides, had no rural counterpart. Victorian conventions nevertheless impinged. The domestic ideology of motherhood, based on the withdrawal of the lady from productive labour, clearly informed the politics of the later *Onward and Upward*. But if domestic service offered 'a reasonably protected path for the socialisation of young women into the dangers of urban life', what happened to those who stepped out of line, and what about unsupervised factory operatives? Valentine discovered that the city with the highest bastardy ratio, Aberdeen, also presented the lowest marriage ratio. If one then considered the excess of females in the population, the explanation seemed crudely self-evident – where there were more women 'at risk', more bastards were likely to be born.[88] Certainly, the presence of women on the labour market was much criticised since it was necessary 'to establish the truth of the maxim that woman cannot be removed from the domestic sphere without danger to her morals'.[89] But an ideal of motherhood based on the withdrawal of the lady from productive labour was limited in its applicability to those who could afford it. Male servants, meanwhile, were scarcely counselled, even though a major factor in the rise of lone parenthood was the relative ease with which men alone could evade the increasingly anachronistic machinery of the church courts as well as the strictures of the welfare agencies.[90]

Instead, the initial loss of social honour felt by the churches, erstwhile regulators of the moral code, expressed itself in the symbolic reaffirmation of a Scriptural family model.[91] In doing this they aligned themselves fully with the Victorian middle class while condemning the immorality of the rural working class. A high price was put upon female chastity. Correspondingly, the more they stressed the role of sex within marriage, the more they had to describe and regulate extramarital forms. Here churchmen, doctors, teachers, sheriffs and journalists were uniquely placed: removed from the immediate sphere of agricultural production they were able to play a mediating role in social relations. Yet, because of their class position, their social fears were often unfounded hyperbole, their 'traditional' family a well-constructed myth of a non-existent past, their social policies still-born.[92] The paradoxical picture of leisured ladies training servants in the virtues of housewifery did not transfer on to workplace conditions, as the collapse of the *Onward and Upward* ideal signifies. Churchmen and philanthropists could do little more than suppress the symptoms since, in large part, the causes of rural illegitimacy lay in material transformations beyond their ken.[93] Housing shortages, the accommodation of single servants, the ritual release of feeing markets – each presented problems of control, either in terms of the breakdown of parental surveillance and 'kindly relations' or via mobility outwith the moral containment of the local peasant community. If the fishing villages were largely self-contained, relations of production inland ensured that farming families were not. Even if the Rural Police succeeded in taming vagrants and the urban reformation in subduing some prostitutes, philanthropists were powerless to apply the voluntary principle to state agencies such as the Poor Law Boards – unmarried mothers preferred to remain on the roads rather than enter the poorhouse.

For Christopher Smout, the failure to turn back the rising tide of illegitimacy and prostitution in Scotland indicated that, *pace* Presbyterianism: 'At root the Victorian moralist, like the Victorian businessman, was an individualist with a privatised vision: it was enough for him if he was pure, even if Sodom and Gomorrah were to be destroyed down the road.'[94] Perhaps, however, the issue was less one of ignoring what was distasteful – or sweeping it under the imperial carpet – than one of misreading the circumstances. Insofar as the labour process could emancipate women from 'traditional' ties, Shorter sees industrialisation and urbanisation as liberating forces. By contrast, Scott and Tilly posit the continuity of traditional values and behaviour in changing circumstances. Thus 'pre-industrial values, rather than a new individualistic ideology, justifies the work of working-class women in the nineteenth century', while illegitimacy arose from the displacement of rural courtship conventions into an urban context. Gradually, however, these older codes were replaced by more instrumental attitudes that saw employment opportunities in terms of 'social and occupational mobility, rather than as a temporary means to earn some money for the family'.[95] Yet, in Scotland, despite massive and sustained urbanisation, bastardy rose outside the cities, and even within them such a characterisation would be unrealistic: social climbing was hardly the lot of the Aberdeen prostitute. For Weeks: 'The key factor seems to have been proletarianisation rather than urbanisation, that is the generalisation of the wage–labour relationship.'[96] The peasant basis of the rural social formation of northeast Scotland rendered it more resistant than most areas to the imposition of capitalist relations of production. Nevertheless, its protracted demise did see the growth of an increasingly landless and potentially homeless labour force.

In his later efforts to combat the 'immorality' of working-class areas of the inner city James Begg came to support relocation schemes. Like those explorers of Darkest England, east of Whitechapel, he remarked: 'We do not see why the crowded districts might not be thinned ... and a great number of the population transferred from the dense centres of our urban districts to the purer air of the country.' He was nevertheless well aware that not all was pure in rural Scotland. Thus, for the farm servant, he maintained the need to 'put an end to roving and unstable habits, by giving them a stake in the country, to produce self-respect'.[97] However, Begg's advocacy of a 'property-owning peasant democracy' worried the Free Church, who were eager to establish that respectable morality was not of necessity dependent on universal home-ownership. The Housing Committee was dissolved in 1867.[98]

The armoury of moral suasion, developed from creditably vast archives of statistical knowledge, was insufficient of itself. By evaluating the failed attempts of an urban bourgeoisie to graft its own ideals on to the social relations of service in the countryside, this paper has exposed a contradiction that lay at the very heart of that system. The displacement of causal hypotheses from the evils of economic exploitation on to a framework of immorality remained crucial to the dominant view, one that emphasised social norms whilst placing minimal reliance upon the law.[99] Thus; although the family, the voluntary principle and self-help gained in currency, church discipline waned while the state regulation of sexual concerns was demonstrably weak. The real problem lay in the fact that their family model was only tenable given specific preconditions – those of the 'property-owning peasant democracy' or those of the urban middle classes – both of which precluded the average farm or domestic servant.

Discussing the 'urbane silence' of creative writing about the nineteenth-century Scottish city, Andrew Noble remarks that contemporary fiction 'scarcely deals with the chronic problem of the age, the new industrial city'. Instead, evasively, the novel is 'suffused with nostalgia for a lost green world...of pietistic peasants...No culture...achieved such a repression of representation of urban life by promulgating such mythical rural alterna-tives.'[100] This paper has aimed to demonstrate that, despite such distancing, the middle class in Scotland had to do a great deal of creative thinking – and writing – in order to sustain the myth of a new moral world in the capitalist future. Such, ironically, was its 'structure of retrospect'.

Acknowledgements
The author wishes to thank the Economic and Social Research Council, under whose auspices some of the research for this paper was carried out, and Aberdeen University Library, for allowing ready access to collected pamphlets in the King, Thomson and Herald Collections, MacBean Room, King's College. I am indebted to Peter Laslett, Philippa Levine, Philip Ogden and Chris Waters for comments on earlier drafts and to my fellow contributors for their constructive remarks.

5

Mobility, the artisan community and popular politics in early nineteenth-century England

HUMPHREY SOUTHALL

The story of nineteenth-century labour is one of movement and migration.[1]

In 1951 Eric Hobsbawm began his classic paper on the tramping artisan with the above words. However, since then little attempt has been made to tell this story: while labour history has moved away from institutional studies of individual unions or hagiographic accounts of 'the movement', these have been replaced, by and large, by studies of the labour process and industrial relations within particular firms, and of communities, exploring the interaction of different working-class institutions.[2] While this shift in focus has unquestionably given greater emphasis to the experience of the individual worker, the geographical focus has moved from the nation to the town or even the individual factory. My central purpose is to argue for broader horizons, not through a desire to move away from the study of individual workers and their life-experience, but precisely because we cannot hope to encompass that experience within studies of single places. In particular, I will argue that the outlook of those workers, principally artisans, who created the central institutions of the modern labour movement was, in crucial ways, conditioned by their experience of mobility, of living and working in a range of places and of possessing a far-flung network of personal contacts.

This essay is, therefore, concerned with the geographical movement of workers in nineteenth-century Britain but not, in a conventional sense, with labour migration. Historical research on movement has been overly influenced by the demographer's or economist's outlook. To the demographer, or population geographer, the objective is to explain the distribution of population between areas, and the movement of individuals is significant insofar as it contributes to changes in that distribution.[3] To the regional economist, the movement of economically active individuals is significant only if it provides expanding industrial areas with a workforce, diffuses the impact of redundancies as the unemployed move away or equalises wages between regions.[4] Lastly, to most social historians and urban historical geographers, migration itself is a largely unexamined process, significant through its creation of culturally distinctive communities of immigrants[5] (see Chapter 3). In these senses, most of the movement discussed here was quite without significance: the travellers moved among their own folk and more often than not ended up back where they started. All that such movement changed was the outlook of the people who moved and of those they met along the way, but this essay argues that such changes were far from insignificant.

The central theme here is the same as Hobsbawm's: the tramping artisan, the skilled worker travelling the country in search of work, on foot and receiving relief from fellow craftsmen. If we wish to study individual experience, we must largely rely on autobiographies, but the remainder of this introduction provides a brief account of the cultural practice and institution of tramping. The early history of travel by unemployed workers seeking employment is beyond my present scope; the subject can be traced back to classical Greece.[6] Further, poor people travelling in hard times would always be in particular need of support from others and were likely to receive it from those who felt a particular affinity to them. Given the significance of occupation in defining affinity, we would expect that much of this aid would come from others in the same occupation; hence the informal support of distressed migrant workers by their craft will have a long history.

What is less clear is when formal systems of aiding travelling members by trade-based societies, within which men from one town possessed a formal entitlement to relief in another, were established. Leeson suggests that by the eighteenth century 'a network of contacts existed between towns', but also that it was 'often informal', meaning reciprocity between independent organisations rather than individuals.[7] The early eighteenth century saw an increasing separation between journeymen and masters, between the yeomanry and the livery within the London companies, linked to the increasing domination of the livery by merchants lacking technical mastery of the craft.[8] By the end of the century, certainly, journeymen's organisations frequently had formal expression as trade-specific friendly societies, meeting at a public house, and while the formal benefits were usually limited to sick pay and a funeral grant, it is clear that in practice they also offered relief to travellers.[9] In a number of cases this provision was mentioned in the rules, but the local society was in no sense part of a larger, national, organisation. The autobiographies discussed below included men who travelled during the eighteenth century, but there are no references to support from trade societies.

Formal financial links between societies, as distinct from 'correspondences', seem generally to have developed at the end of the eighteenth century, possibly reflecting easier travel and greater mobility; among woollen workers, for example, the 'Brief Institution' linked Yorkshire with the west of England from 1799 and was sustained by tramping.[10] The extent of such formal links is difficult to estimate due to the Combination Acts as well as to restrictions on multi-branch societies by the Registrar of Friendly Societies, but by the repeal of the Acts in the mid-1820s Leeson estimates that links existed in at least twenty-eight trades.[11] As these links developed, two crucial elements appeared. The first was some system to prevent fraud, to stop men who were not members of the trade claiming relief; this usually meant an identity card, or a ticket handed out at one branch entitling the traveller to relief at the next. The second was the equalisation of funds, by which branches whose end-of-year balance of funds was above average remitted money to those with below average funds under the direction of some central branch. Equalisation effectively pooled the funds of the entire organisation without creating a single central fund vulnerable to fraud or expropriation and was the crucial innovation marking the move from local societies providing reciprocal benefits to a single national body. Precautions against fraud generally came first and in some trades the first national organisations were specifically limited to policing the travelling system.[12]

The formal travelling system was essentially a creation of the early nineteenth century and was only fully developed in the 1830s and 1840s, precisely the period when the impact of the railways and the increasing integration of the national economy began to undermine the reasons for its existence. However, it built on a long tradition of mutual aid within the artisan community and, as the accounts of travellers show, even in the 1840s men depended in practice on a combination of formal relief by their society and informal relief from fellow craftsmen as well as relatives and friends. The essence of the formal system was that members in search of work were provided with a list of the public houses where the other branches met, together with a card or 'travelling certificate' proving their entitlement to benefit. On arrival at another branch they would present this card, be informed whether work was available there and, if forced to travel on, be provided with relief. Relief took the form of either customary assistance, meaning a meal, a pint of beer and a bed for the night, or a sum of money related to either the number of days since last relieved or the distance travelled.[13]

Those unions who could afford it replaced this system by conventional unemployment pay and the payment of rail fares following the economic crisis of 1848, whose impact was national and hence inescapable however far a man travelled. In those trades that could afford nothing else, such as many building crafts, or those for whom cyclical unemployment was a lesser problem, such as the printers and the tailors, tramping lingered on into the 1880s and 1890s or even the present century.[14] However, by then the tramp had become increasingly marginalised within the unions, and even within the artisan community the word had acquired its modern connotations.[15] In

institutional terms the importance of the travelling system was threefold:
first, it represented the first systematic attempt by a new industrial workforce
to come to terms with economic distress manifesting itself as complete
unemployment rather than short-time or intermittent working; secondly, it
created national unions of trades when there was little need for wide
organisation for bargaining with employers and so was crucial to the
emergence of a national labour movement; thirdly, it grew out of, and gave
concrete form to, a long tradition of geographical mobility and aid to the
traveller within artisan culture.

Travellers' tales

Our existing knowledge of the travelling system and its consequences comes
mainly from the study of trade union rulebooks and accounts by outsiders;
hence it is seen very much as something done by trade unions and its
consequences are primarily institutional. Our concern here, however, is with
the individual travellers themselves and quite different sources are needed:
the following account is based on a reading of fifty-two autobiographies of
people who went 'tramping', drawn from a major bibliography. Here
tramping means setting out, usually on foot, in search of work, and the
majority make no mention of trade-society assistance. Only one was a
woman, and she a rather unlikely tramp; hence the references here to
'men'.[16] Lastly, the term 'tramp' lacked its modern associations; men
happily referred to themselves as tramps.

While this source in some ways takes us closer to individual experience
than any other, it must be interpreted with some care. First, the men covered
are self-selected and a quite unrepresentative fraction went on to achieve
prominence as industrialists, union leaders, politicians, poets or even wood
carvers. A number were active Chartists, but of these Dorothy Thompson
comments:

A handful of autobiographies provide valuable historical sources, but they have also
distorted the picture by their very existence. A few Chartists who gained prominence
in later life as literary or political figures have been the subject of biographies or
biographical sketches, but here again, the type of person of whom it has been possible
to make such a study has usually been an untypical member of the movement.[17]

We therefore cannot take the collective experience of these men as, in itself,
typical; we must rather identify the norm by discovering which of their
experiences they saw as typical and which deserving of special comment.

A second problem is that we are generally dealing not with a daily record
set down at the time but with an account written in old age, frequently with
a particular aim in view. Eric Hobsbawm's comment on oral history applies
just as strongly here:

We shall never make adequate use of oral history until we work out what can go wrong in memory with as much care as we now know what can go wrong in transmitting manuscripts by manual copying.[18]

The majority of the authors are concerned with demonstrating how their lives were transformed by socialism, religion, their own efforts or the pledge, and there is therefore some tendency to exaggerate the humbleness, depravity or hardship of their early years. This is taken to an often farcical extreme in the autobiographies written as temperance tracts; an example is Miles Watkins, who claimed to have lost £40 – 50,000 through 'the accursed and brutalizing drink' between his birth as the impoverished fifth son of a pig-dealer in 1772 and 1841, without ever quite explaining how he earned such a remarkable sum in the first place.[19] The autobiographies of Chartists are perhaps the least affected by this problem, as most were born into solid artisan families and rose only slightly if at all through their political activities.

The autobiographies read were those that referred to 'travelling', so it is unsurprising that almost all authors made substantial moves. The reading of all extant autobiographies would be a major undertaking and might give more direct evidence of levels of mobility. It is worth quoting the comments of the compilers of the bibliography:

For that minority who [became] skilled engineers, printers, cabinet-makers, stonemasons and the like... occupational changes were not so frequent, though the biographies illustrate a high degree of geographical mobility in these trades, especially for unmarried men...

Periods of unemployment and under-employment were within the experience of a high proportion of the autobiographers here represented. This helps to explain the frequency of tramping by all grades of labour, not just the skilled artisan. Sometimes the motives were simply variety and interest, sometimes to gain wider experience of new machines and technologies but often tramping was a search for work in a period of unemployment or when an old craft was decaying.[20]

However, in general what we can hope to learn from the autobiographies is not the quantity of movement but the quality: the why and the how, the attitudes of the traveller to his journeys and the consequences for him and others.

The degree of detail provided by the authors varies greatly, some giving precise itineraries, others vague descriptions such as 'for sixteen months I tramped through the principal towns of Middlesex, Lancashire, and Yorkshire'[21] or, unhelpfully, 'I need not follow my wanderings for some years, as my life at that time was of the ordinary kind'.[22] For this reason, the actual operation of the travelling system is better documented from more systematic information, and the following account draws on a reconstruction of movements within one particular artisan union, the Steam Engine Makers' Society (SEM), using information in its annual reports for the 1835–46

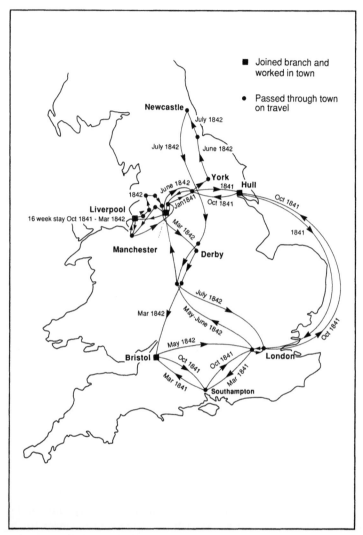

5.1   Movements of Thomas Watson, 1839–1842
   (*Source: Steam Engine Makers' Society* 'Annual Reports')

period. Systematic analysis of these data is presented elsewhere,[23] but two examples give some flavour of the individual histories that lie behind the overall level of movement; the men chosen are both extremes, as obviously many men made no moves at all.

Thomas Watson (Figure 5.1) was notable for travelling furthest in a single financial year, 1,332 miles in 1841–2. He belonged only briefly, joining Manchester branch in 1839–40. In January 1841, he tramped to Hull and

worked there until March; he continued via London and Southampton to
Bristol, working there until October. He then returned via Southampton,
London, Hull and Leeds to the northwest and joined the Liverpool branch.
He received sixteen weeks' sick pay there in March 1842, suggesting he was
ill all winter, and then left on a complicated journey that took him via
Blackburn, Derby and Birmingham to Bristol, where he stayed a couple of
months without rejoining the branch. He then went to London, Tipton,
Manchester, around the northwest and then north to Newcastle. However,
he did not stay there but immediately returned via Ripley and Tipton to
London, being relieved there on 27 July 1842. The last reference to him is a
payment in East London, where he was detained 'on account of his
Travelling Certificate', suggesting that he was then expelled for irregularities.

Ellis Rowland (Figure 5.2) left and returned to Manchester four times. He
was first listed at Manchester in 1835–6, joined Liverpool in that year, but
returned to Manchester during 1836–7. In April 1838 he tramped via
Liverpool to Hawarden and worked there, but returned to Manchester in
August. During the following financial year he joined Kingswinford branch
in the West Midlands and stayed there until November 1840, when he
tramped back to Manchester. His final move began in August 1842, when a
sequence of tramping payments trace his journey via Newton-le-Willows,
Liverpool, Hawarden and Tipton to Birmingham. He returned to Manchester in February 1843, a tour of the northwest suggesting that he was
returning not because there was now work in Manchester but because the job
in Birmingham had ended. He then remained in Manchester until the end of
the study period.

Although these are both extremes, many other members of the SEM
moved almost as much. However, we learn little of their actual experience
and for this we must turn to the autobiographies. Space alone prevents the
detailed presentation of many careers, but one example is worth presenting
as a whole: Robert Gammage's life is of particular interest because of the
relationship between his movements during the 1840s and Chartism.[24] His
reminiscences are far from a complete account of his life but provide a fairly
detailed account of his movements in the early 1840s; the major journeys are
mapped in Figure 5.3. He was born in Northampton in 1820 or 1821 and
served an informal apprenticeship as a coach trimmer. He first left
Northampton in February 1840, having spoken at several political meetings
in the area during 1839, and lost his job as a result. At this stage he was
simply a young artisan, albeit politically active, seeking work and using the
houses of the United Kingdom Society of Coachmakers as well as a
remarkable array of personal contacts.

Leaving Northampton on 6 February, he went first to Newport Pagnell
and stayed at the clubhouse. He went next to Bedford, detoured to Ampthill,
went on to Huntingdon, where he stayed with the mother of a friend, and

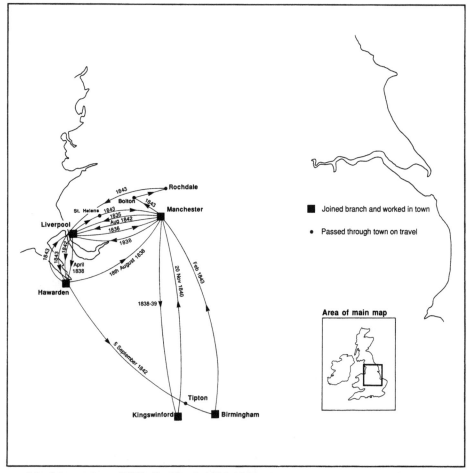

Joined branch and worked in town

Passed through town on travel

Area of main map

5.2   Movements of Ellis Rowland, 1835–1846
     (*Source: Steam Engine Makers' Society* 'Annual Reports')

then on to Cambridge and Hertford. There he stayed with friends and made
a visit to a cousin in Hatfield whom he had not seen since childhood. In
London he stayed initially at the clubhouse but then lodged for six weeks
with an uncle and aunt. He moved on to Sevenoaks and Maidstone, where
he had another friend from Northampton, to Tonbridge, Lewes and
Brighton. From Brighton he made a day trip to Balcombe to visit a half-sister
and then continued along the coast to Chichester, Fareham and Southampton. There he heard work was available in Portsmouth, but he was
rejected as too inexperienced. He travelled on to Salisbury and Devizes,
where he met another friend from Northampton, to Bath and finally to
Sherborne in Dorset, travelling the thirty-seven miles in one day:

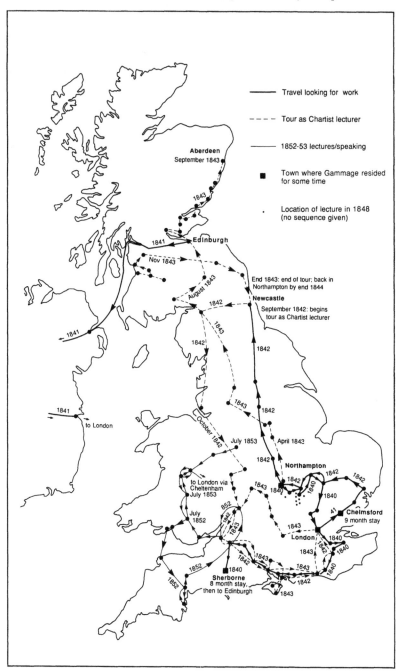

5.3   Movements of Robert Gammage, 1840–1853
      (*Source: autobiography*)

I arrived in the dusk of evening, having made an early start, and of course went to the inevitable clubhouse. I was informed by the secretary that there was an opening for me if I chose to stay.[25]

Although he knew no one in Sherborne, he discovered that Jerry Haggerty, whom he had become friendly with through meeting him on tramp in both Lewes and Chichester, was working six miles away in Yeovil, 'and I, of course, paid him a visit'. He makes it clear that both Devizes and Sherborne were very hostile to Chartism,[26] although his employer did not penalise him:

Every Sunday morning [in Sherborne] I received by post the Northern Star... Many of my friends were eager to get a look at [it], and I gratified them as best I could. Those that read it did not think that Chartism was so bad as it had often been presented.[27]

Unfortunately, work became scarce in Sherborne and after eight months he had to set out again:

I became involved in a long and arduous search for work, and during that time felt the full force of my remark to Mr Hill [his employer in Sherborne] on the value of a trade society to support men when seeking employment. I travelled no less than 1,400 miles in different parts of England, Wales, Scotland, and Ireland before I again obtained work.[28]

His precise route is unclear, although it took in Halifax and Bradford, and he was in Edinburgh for New Year's Day 1841. From Glasgow he took a steamer to Belfast and later visited Dublin before returning to London and finally finding work in Chelmsford.

In late 1841, after nine months, he was dismissed for political activities and tramped back via Ipswich and Cambridge to Northampton, where he found work for three months. Travelling again, he passed through Leicester, which he describes as 'in a constant ferment' due to 'intense distress'. In Sheffield he met the Chartist leader George Julian Harney and in Leeds worked for seven weeks and addressed meetings there and in the surrounding townships. He stopped briefly in Harrogate, where he had an introduction from his employer in Sherborne to a coach trimmer who had moved there from Dorset, and he finally arrived in Newcastle in September 1842. It is clear that on his travels since Chelmsford he had become increasingly active as a speaker, and in Newcastle he was advised to take up lecturing as a Chartist orator regularly.

During the remainder of 1842 and 1843 he travelled very extensively, making a sweep through the northwest, the Midlands and the south to arrive in London for Christmas 1842. He then went to the southwest and south in the early spring of 1843, to the East Midlands, Yorkshire and Lancashire in the early summer, to Scotland in the autumn and finally back to Newcastle in November 1843. He had developed links with radicals in the northeast and remained there for several months in early 1844, but by the end of the year he had returned to Northampton. In 1845 an argument with the Northampton Chartists led him to move to Stony Stratford, working first as a hawker and then as a shoemaker. In 1848 he became involved in the upsurge of popular agitation, lost his job and moved again to Buckingham. In this period he lectured in many towns in the area.

A further period of inactivity ended in 1852, when he was elected on to the executive of the National Chartist Association. Following this, he again travelled extensively as a speaker and fragments are again shown in Figure 5.3. The reminiscences include detailed descriptions of some of these journeys, mentioning the 'fine Cathedral' in Exeter, the 'beautiful River Avon' and so on.[29] He was a candidate at Cheltenham in the General Election of 1852, an incident also remembered by William Adams, then a young activist in the town.[30] In 1854 he fell out with Ernest Jones and was not re-elected on to the Chartist executive. He then moved to Newcastle and Sunderland, where he spent the years 1854–87, qualifying as a doctor and working mainly for friendly societies. In 1887 he was forced to retire because of ill health and returned to Northampton. He died there in January 1888 after falling from a tram.

Robert Gammage's life has been described at some length as it demonstrates a number of very important points. First, even as a young coachmaker, not yet twenty, he could draw on an astonishing range of friends, relatives and former work-mates scattered over southern England; without these, and his trade union membership, his travels would have been vastly more difficult. Secondly, in the course of his travels Gammage both spread the gospel of Chartism, bringing the *Northern Star* to Sherborne, and was himself increasingly drawn into the movement. His later career as an orator reminds us of the crucial role of often obscure travelling 'agitators' in knitting together a national movement that lacked access to modern methods of communication. Thirdly, despite much movement some concept of 'home' seems to recur: both Northampton and the northeast. Finally, this career makes clear why the census and similar sources may be so profoundly misleading: conventional demographic sources would tell us that Robert Gammage was born in Northampton, was living there, or nearby in Buckingham, in 1851 and eventually died there in 1888. It is hard to imagine a more misleading summary of a career.

Starting out

The examples given in the previous section raise obvious questions about what was typical: Who travelled and how much? At what stage in their lives, and at what state of the business cycle? Above all, why did men start out on the road? The existence of travelling relief within the artisan trades obviously made it easier for them to travel, but it is very difficult to demonstrate that the skilled were more mobile than the unskilled.[31] This is particularly true if we measure mobility as total distance travelled over a period, rather than as distance from an origin such as birthplace, and the strongest argument is on a priori grounds: the economic rationale for geographical movement within a segment of the labour market was either to gain a range of experience of different workplaces or to locate a job vacancy through a search process. The more specialised the employment, the greater the distance that would normally have to be travelled to visit a given number of workplaces, unless the occupation was highly localised. Therefore, farm labourers moved around small circuits and textile workers generally remained within their districts, but artisans frequently needed to travel around the country as a whole to gain wide experience or simply to find a vacancy.

Within the records of the SEM, evidence on levels of mobility is much clearer, and the most startling feature is simply how much movement there was. Over the 1835–46 period, on average 13 per cent of the membership left their branch for another town each year, rising slightly in the early phases of the 1842 depression, and over 7 per cent crossed a county boundary or went abroad. Of the 896 members of the SEM who belonged for six years or more between 1835 and 1846, 49 per cent made no moves, 15 per cent moved from one town to another within a county at least once, 30 per cent changed counties and 6 per cent went abroad; a man who moved at all made an average of 2.5 moves. Looking at sequences of movement, return moves were common: 154 of the six-year members made at least one move whose destination was the branch at which they were first recorded; this represents 17 per cent of the group as a whole, 33 per cent of those who moved at least once and 52 per cent of those who moved at least twice. Thus in broad terms over half of those who made sufficient moves for a return move to be included actually made such a move. Linkage between membership lists in the rule books of the Newcastle and County United Tanners shows that 12 per cent of the 505 members listed in 1826 were named in a different town in 1830. Restricting ourselves to those who can be identified in both lists, 31 per cent had moved town by 1830, suggesting levels of mobility comparable to the SEM. In general, these statistics suggest that the typical member of a tramping society made at least one major inter-urban move during his career.[32]

Within the 1835–46 period, levels of tramping in the SEM were strongly concentrated into depression years, when both the proportion of members going on tramp and the average distance they travelled rose substantially; in the worst year of the depression, 1841–2, 227 members out of 1,202 (19 per cent) received at least one travelling payment; the average distance covered, chiefly by walking, by these 227 was 334 miles, with 127 travelling over 200 miles and 6 over 1,000. However, this may partly reflect rules that limited relief to men who had lost their job rather than simply quit.

Table 5.1. *Mobility between 1835 and 1846, by date of birth*
(SEM members of 6 years' membership or more, of known age)

Date of birth	Number of members	Number ever moving	Nature of greatest move			Number of moves		
			Within county	Inter-county	Over-seas	1	2–3	> 3
1815–1824	65	33 (50.8)	11 (17)	17 (26)	5 (8)	11 (17)	17 (26)	5 (8)
1805–1814	97	45 (46.4)	17 (18)	24 (25)	4 (4)	16 (17)	18 (19)	11 (11)
1795–1804	64	26 (40.6)	11 (17)	14 (22)	1 (2)	9 (14)	14 (22)	3 (5)
Pre-1795	64	20 (31.3)	10 (16)	10 (16)	0 (0)	10 (16)	7 (11)	3 (5)

(Bracketed figures are percentages of total in age group)

Table 5.2. *Levels of tramping by age group*

	1839–1840				1842–1843			
Age	Number of men	Ever tramping No.	%	Mean mileage per tramp	Number of men	Ever tramping No.	%	Mean mileage per tramp
Under 30	100	14	14	194.9	122	39	32	383.6
30–39	90	10	11	77.5	98	21	27	274.7
40–49	53	7	13	104.9	62	13	21	234.5
50–59	28	5	18	149.4	35	5	14	63.8
Over 60	8	0	0	—	13	3	23	69.1

(Sources: Steam Engine Makers' Society, 'Annual Reports'. Ages from the membership register of the Amalgamated Society of Engineers (many SEM members transferred to the ASE in 1851); the funeral records of the SEM, which were searched up to 1905; and death certificates for men whose funerals were recorded by the SEM without an age being given)

Finally, the SEM records provide us with some evidence on the degree to which tramping was concentrated into certain stages of the life-cycle. Tables 5.1 and 5.2 examine, for members whose ages could be located, the relationship between the stage of the life-cycle and, respectively, branch transfers and tramping activity. Note the relative shortage of really old men; even someone born in 1790 was only fifty in 1840, in the middle of the study period. It is clear from Table 5.1 that young men were more mobile, but not drastically so; much of the difference reflects a greater propensity to emigrate, although the sample is small. There are rather larger variations in the use of the tramping system by different ages. In 1839–40, in the early stages of the depression, when economic conditions were worst in the north-western heartland of the SEM and there were real advantages to going south, all ages seem equally likely to tramp, although there it seems that men in their forties travelled least far. However, in the depths of the depression in 1842–3, the mean mileages travelled show that it was principally the young men who were tramping the country in a generally fruitless search for work.

These findings are important because the autobiographies talk about tramping mainly as something done in youth, at the end of apprenticeship. If, as the statistics suggest, tramping was somewhat biased towards the young but in no sense limited to them, there are two explanations for the autobiographers' emphasis. First, many of the authors, and particularly those writing lengthy accounts, had made drastic changes in social status by late middle age, when 'ordinary' artisans were seemingly taking to the road again: Gammage was a doctor, Henry Broadhurst a government minister, William Fairbairn a leading engineer and so on. Secondly, it may be that for the older men tramping was an unfortunate necessity, but for the young it was also an emancipating experience and so would be a central element in a retrospective account of how a man came to be what he was. This is just one point where we must remember that an autobiography can seldom be read as a day-by-day chronicle.

The two alternative explanations of tramping, as a response to the economy and as a stage in the life-cycle, can be found in individual accounts. Much the commonest reason given was simple lack of work in the local area; Henry Broadhurst described this in a literary fashion:

Just as when you look into a kaleidoscope ... you give the instrument a turn, and the pieces of glass fall away into new positions ... so in the sphere of labour the changes and chances of commercial life and the caprices of fashion keep a large army of working men in a state of motion, sometimes over short distances, sometimes from the southern counties to the western, or the eastern to the northern.[33]

One feature that emerges very clearly was individuals' awareness of the economic forces that moulded their lives, and particularly the business cycle: 'business was flat'; 'work was rather slack'; 'trade was at a low ebb'; 'there was a great fall in the trade'; and so on.[34]

However, other men talk explicitly of wishing to see the country, to learn other men's methods of working, or to emulate travellers' tales of drinking and 'getting agate with young women'.[35] As one young engineer, Thomas Wood of Keighley, put it:

I thought I was deficient in my trade, though I learned all I could learn there. But I heard about new tools, new machines, and new ways of working. I could never hope to see them in our shop, and if I was to learn, and improve, I would do so now before I either married or thought of it. So one Friday in the summer of 1845 I left my old master for good and ever.[36]

In addition, many men started travelling at the end of their apprenticeships, when their masters were unwilling to keep them on at higher wages as journeymen, and it would seem that this type of 'unemployment' was often expected.[37] Thomas Wright commented that 'the majority of [the tramps in a London clubhouse] are generally young fellows who have come up from provincial towns on completing their apprenticeship'.[38]

Some men travelled to get away from troubles. Several were facing victimisation for their political or trade union activities, including several active in Chartism.[39] George Herbert, a Banbury shoemaker, was sent to work on the coast because of his health,[40] while Miles Watkins repeatedly left Cheltenham to escape imprisonment for debt.[41] However, a common reason for running away was to escape either cruel parents[42] or, for apprentices, a harsh master;[43] both of these would seem to fit the notion of travelling as a rite of passage.

Life on the road

Attitudes to the actual process of travelling seem to vary, although this may reflect the aims of a particular autobiography rather than what a man actually felt at the time. Some, like Henry Broadhurst, stress the discomforts of travel:

The frost, gripping my sodden boots, had turned the leather to the consistency of cast iron. With great difficulty I got them on; but when I came to walk, their unyielding surface chafed my feet sorely, reopening the wounds which had all but healed. Walking under these circumstances keenly tormented me; but in spite of all I managed to cover the ground between Brighton and Tunbridge Wells in ten hours.[44]

However, many of the authors talk of the pleasures of the countryside and went out of their way to visit famous sites, such as Shakespeare's birthplace.[45] Robert Gammage, talking about walking from Brighton to Worthing, comments that 'everyone who knows the coast is aware that the walk from one town to another is most enjoyable'.[46]

Although there are occasional references to problems with understanding local dialects,[47] being expected to pay 'footings' all over again in a new town,[48] or simple hostility to strangers,[49] most men report a friendly reception as they travel. Travel was eased by going with a mate, sometimes by arrangement from the start,[50] sometimes met by chance along the way.[51]

The freemasonry of the road is one of the charms of tramping. Every tramp chums up with every other tramp, just as if he had known him from boyhood.[52]

This freemasonry was not limited to members of the same trade:

Sometimes it would be a bricklayer, sometimes a tanner, and sometimes an engineer. If our goal was in the same town or village, we would journey together as long as our ways lay in common.[53]

Note that the occupations mentioned, and the author's trade of stonemason, were all artisan crafts.

'Tramping' meant walking, although advantage was taken of other modes when available and affordable:

If the poor rode at all in pre-railway times, they had to ride in wagons, living and sleeping in them for days and nights even on comparatively short journeys. I remember them well, those great lumbering wagons, as big as haystacks, covered with tarpaulin, the wheels broad enough in the tyre to span a ditch, the six or eight horses apparently strong enough to move a mountain. Coaches were for those who could afford a more rapid transport.[54]

We easily forget the importance of coastal passenger boats, which continued to be used well into the railway age as a cheaper alternative, so that even in 1895 Fred Bower, going from Edinburgh to London on union business, found himself on a blazing ship off Clacton.[55] The railways were a major factor in the decline of tramping, but in the 1830s and 1840s they were only starting to assist the better-off craftsman, as illustrated by George Herbert's account of his wedding in 1837:

At this time, the Queen's coronation came on...I thought this would be a good time for me to get away for a few days, so I started off [from Banbury] to Dover to get married and bring my wife home. So I started on the Sunday morning and walked to Buckingham and from there to Stony Stratford. I though I could there meet one of the coaches which came through as the railroad was not finished, but was partly made at one end and partly at the other. At the upper end a temporary station was made at Denbigh Hall [near Bletchley].[56]

He found a perch on the last coach of the day, but then someone tipped his hat off into the road and after retrieving it he had to walk on to Bletchley. There he caught the last train to London. On the Monday he took the boat to Dover, the banns were posted on the Tuesday, he was married on Wednesday, returned to London on Thursday and then on Friday:

We started home and as I wished to be as economical as possible with my finances we booked for Banbury on one of Rudkin's fly-vans that used to run to London to Newgate Street with meat and butter. This took a night and a day to get home, but coach hire at that time was very expensive [costing over a pound]. I patronised Rudkin's van several times after, and other small tradesmen situated in similar circumstances to myself used to do the same. I think the fare was five shillings for the single journey.

Well, myself and wife got home all right, and I was ready for work on the following Monday.[57]

Herbert is describing how an independent craftsman in funds would make a long journey and it is unsurprising that unemployed men simply walked; with the exception of boat fares, this was all that most trade societies provided for. Note also that Herbert had met his wife when he went to work in Dover as an apprentice, so that this man, who was born in Banbury in 1814, was living there in 1851 and died there in 1902, had a wife born elsewhere. Such a pattern was probably not uncommon.

A striking feature of most of the accounts is the support that travellers could call on in the towns they passed through. Examples have already been given in the career of Robert Gammage, but the full range is made clear by other authors. Many could call on relatives for assistance and the majority would seem to have had a sister, or an uncle, or at least the cousin of any employer who could provide a bed in London.[58] Frequently on their travels the authors met friends or former work-mates, who again offered assistance.[59] It would generally be members of social groups that were most mobile who would be most likely to have a widely scattered networks of relatives and friends, but even they could not rely on knowing someone in every town. Here affinity groups, both formal and informal, become crucial, and while co-religionists may have been important to active non-conformists,[60] the dominant informal affinity group were workmen of the same craft and the main institutional source of support was trade societies. As Thomas Wright put it:

The majority of working men have for a man on the road that fellow-feeling that makes us wondrous kind. Most of them either have been, or know that they may at any time have to go, on tramp, and so they adopt for their motto, 'Be to a friend in distress like a brother', and receive a travelling fellow-craftsman in all cases with brotherly kindness.[61]

A fine example of this is given by Henry Broadhurst, who tramped from Norwich in the depressed winter of 1858–9 and made for Portsmouth because he had heard there was work for masons on the Royal Victoria Hospital at Netley. However, 'the same idea had attracted many others out of work' and he found the road 'swarming with men imbued with the hope of finding employment on the government building'. Broadhurst was by then in poor health from his travels, but he was helped by some fellow-masons who had joined the militia:

They lent a ready ear to my necessities, and at their suggestion I entered my name on the sick-list of my trades-union, and obtained a week's lodging in its headquarters in [Portsmouth].[62]

They smuggled him in to the barracks to eat and after this week's rest he was fit enough to travel on to Brighton. Several similar examples could be given.[63]

However, members of trade societies had a more certain source of support; as Thomas Wright put it:

the class who are most frequently found on tramp are the mechanics who are members of trade unions. For them, the road is deprived of half its terrors and inconveniences... In almost every town they have their clubhouse, at which they will perhaps meet some old mate, and at all times find fellow-unionists and brother craftsmen who will receive them in good fellowship, and furnish them with reliable information as to the state of trade and the chance of obtaining work in the town.[64]

The greatest advantage of the clubhouse as a source of relief was its reliability, but the standard of accommodation would often seem to have been quite high:

[The stonemasons' union] had relieving stations in nearly every town, generally situated in one of the smaller public houses. Two of the local masons are appointed to act as relieving-officer and bed-inspector. The duty of the latter is to see that the beds are kept clean, in good condition, and well aired, and the accommodation is much better than might be expected.[65]

Accommodation was clearly far superior to the common lodging houses that tramps were increasingly forced to use later in the century,[66] or if they were not members of a union,[67] and was arguably crucial to men retaining their self-esteem and a 'respectable' appearance. The greatest defect of the system for the traveller were the gaps in the network; in southern England there was often as much as forty or fifty miles between stations, and then 'the traveller's life became a hard one'.[68]

One final example illustrates the use of national societies as a ready-made contact network. In the 1830s John Buckmaster, unhappy with his apprenticeship, decided to run away from Uxbridge but was handicapped by never having been more than thirty miles from the Chilterns, 'and a journey of one hundred miles into unknown country required some consideration'.[69] He was particularly concerned that if he ran away with his vital but heavy tools, his master would soon catch him on horseback. His solution came through his membership of the Rechabites friendly society, which he describes as 'a kind of secret society with a password'.[70] He examined their list of lodges, wrote to the secretary of the lodge at Andover and asked him to take delivery of his tools, which he sent off secretly by carrier a week before his own departure on foot.[71] This use of a friendly society is unique; in general, reliance was on trade unions.

One fundamental need of the traveller was food and shelter, but the other was information about employment. Some men were following regular, often seasonal routes, and knew from past experience that work was likely to be available. Seasonal migrations were present in a surprising range of trades. In shoemaking, George Herbert benefited as an employer from his proximity to Oxford:

I used to encourage the tramps to come through [Banbury]...They used to go to Oxford for the term and when that was up they used to come down to me, so you see I always had a lot of superior workmen.[72]

Similarly, when Paul Evett worked as a compositor in Warwick he found himself joined by members of 'the travelling fraternity' who knew that at that time of year there was extra work because of printing the electoral roll and the County Directory.[73] Some seasonal movement involved a change in occupation: men from the gas works going to the brickfields in summer[74] or even tramp printers early in this century painting lampposts in summer;[75] unskilled men might find work harvesting.[76] Lastly, some annual movements reflected labour markets where men were hired for one year at a time; the best-known example was that of farm labourers changing employers at annual hiring fairs and moving around geographically relatively constrained circuits,[77] but something similar operated among the Durham miners:

Little more than 20 years ago, when the pitmen were bound to collieries from year to year, there existed a rather peculiar system in Durham county. Those miners who were dissatisfied with one colliery would get bound at another, so that shifting chattels and family became necessary. In this way, thousands of families would change district annually, and in April and May the whole mining districts were alive from side to centre with vehicles from big waggons to 'cuddy' carts.[78]

Those making less regular movements were sometimes going to a specific job, having been sent for or carrying a letter of introduction.[79] In other cases, they had written applying for jobs, but in these instances they were increasingly likely to travel by train.[80] However, many were genuinely travelling as part of a search for work. Sometimes casual work would be easy to find, so that William Farish happily set out to Glasgow from Carlisle in 1836 'without provision of any kind' and 'minus even a solitary copper', knowing that 'by tarrying a day or two at one place, and a day or two at another, the journey could be broken and the wants...supplied',[81] but often the assistance of strangers was vital. One source was other tramps who were met on the way or who passed through prior to a man's departure:

A new arrival at a tramp's bar would exchange information with the 'resident' and give tips about the places where he had worked on his itinerary.[82]

As men travelled they heard about possible jobs from people met on the way, sometimes from casual conversations on the roadside.[83] However, for the member of a trade union the most important source of information was the clubhouse, especially if he had previously 'worked mates' with any of the local members:

If any of them have influence in the establishment in which they are employed, they exert it to the utmost in trying to get work for their old mate... Trade must be very dull indeed if in the large towns a man who has friends in the trade... does not get into employment in the course of a few weeks.[84]

Mobility and consciousness

Given the wider concerns of this essay, the central question is not why or how men travelled but the effect their travels had on them and on the people they met. First, despite the importance of economic factors in setting men on the road, it is clear that for young men the system played a vital role in socialising them into the wider artisan community. Further, tramps, particularly older men, were essential to passing on the traditions of the trade, especially in more isolated areas:

The tramps were the principal medium of trade news and were as important to youngsters on the outskirts of the trade as were the reporters of 250 BC who used to report in a vernacular Sanscrit in Delhi.[85]

At William Marcroft's initiation as a journeyman, it was a visiting mechanic on tramp he remembers as standing up to give a few words of advice.[86] Without travelling themselves or meeting travelling fellow-craftsmen, members of specialised occupational groups scattered over the country would have developed little sense of a common identity. This common identity, in turn, provided them with a framework of support within which they could easily move around; the system therefore worked to reinforce itself.

Another effect of travelling was to create a sense of independence, which arguably grew in importance as men became, in economic terms, more proletarianised: they might be dependent on having an employer for the means of subsistence, but they always had the option of moving on.

One recurring theme is the importance of a man's tools, ownership being the distinguishing mark of a craftsman. Men frequently went on travel with their tools, despite their considerable weight, and make specific mention of buying or repairing their tools when they see a period of travelling looming.[87] On the other hand, the main factor that tied men down was a wife and family, and marriage appears as a crucial transition, ending a period of mobility that started with the completion of a man's indenture. Once married, travel became far more of an imposition, although a wife might be sent home to her mother's to free a man to travel;[88] examples of married couples travelling together are rare.[89] The analysis of the Steam Engine

Makers' Society suggests that in later life men again took to the road more easily, possibly following the death of a wife and the independence of children, but there is little evidence of this among the autobiographers, possibly because many had by then made major and unusual changes in status.

The role of travelling in the development of unionism is not central to this essay and in fact the autobiographers say little about it, perhaps because the link between the system and national unions was self-evident. The principal exception was a coal miner, Edward Rymer, whose experience of travel between as well as within coalfields seems to have been unusual; he speaks explicitly of 'missionary work'.[90] Although some of the autobiographers became prominent in their unions,[91] the crucial question is rather whether trade union leaders generally had a prior history of mobility. A systematic investigation of this is beyond my present scope, but examples of prominent artisan leaders can be given. The two principal architects of the Amalgamated Engineers were William Allan, born in 1813 in Ulster of Scottish parents, brought up in Scotland and apprenticed in Glasgow, who moved to Liverpool in 1835 and subsequently to Crewe;[92] and William Newton, born in Congleton, Cheshire, in 1822, who was apprenticed at Etruria and joined the Hanley branch of the Journeymen Steam Engine Makers in 1840 but soon moved to London, where he immediately became active in union affairs.[93] The second largest national union in mid-Victorian Britain was the Amalgamated Society of Carpenters and Joiners, and their first leader was Robert Applegarth; he was born in Hull in 1834 and informally apprenticed there, moved to Sheffield in 1852 and married there in 1855. However, although Sheffield was where he rose to prominence in the union, after his marriage he moved to America and had varied experiences there, only returning in 1858 because his wife's health prevented her from joining him.[94] That three such crucial figures in the history of artisan unionism all made major moves prior to the start of their union careers suggests that this broadening of experience may have been an important catalyst.

The autobiographies provide far more evidence on the relationship between mobility and political activism and here we can supplement them from other sources. Popular political movements of the early nineteenth century were hampered by restricted access to what national media existed and by constraints on national organisations through the Seditious Societies Act. They were diffused and sustained by travellers: 'Chartism originated as a platform movement' and it was made national by the work of travelling speakers.[95] Although great emphasis has been given to corresponding societies linking areas,[96] the writing of letters is seldom the origin of a relationship even if it subsequently sustains one; by and large, the first contacts were made not by epistles but by missionaries. Some of these missionaries of the radical movement were and are well known: for example,

Major Cartwright, a veteran agitator active from the 1790s to the 1810s, of whom Thompson says:

It is difficult to over-state the importance of Cartwright's evangelizing tours of 1812, 1813, and 1815. For fifteen years the pockets of parliamentary reformers throughout the country had been without a national leadership ... It was the inflexible Major, now over seventy, who decided to enter the Luddite counties ... In his tour of 1812, he held meetings at Leicester, Loughborough, Manchester, Sheffield, Halifax, Liverpool, and Nottingham. In January and February 1813 he undertook a second tour, holding meetings at thirty-five places in the Midlands, north and west in less than thirty days.[97]

Soon after this, Henry Hunt and William Cobbett were almost as active; 'their progresses resembled those of the most popular Royalty, and their appearances those of a prima donna'.[98]

In the Chartist period the same was true of Feargus O'Connor, who was above all a leader of the people, a demagogue:

O'Connor was the national figure whose visits were the occasion to organise massive demonstrations, to exploit every theatrical device ... He played on this function of the figurehead, dramatising his personality and using the demagogic rhetoric of the sacrificial leader.[99]

Figure 5.4 shows the towns where O'Connor spoke in just two years, 1838 and 1839, and some of his tours.[100] Although O'Connor travelled by train if a line had been completed, and by 'chaise and four' otherwise, such journeys still required extraordinary physical stamina; a total of 147 public appearances are plotted, while in September 1838 he told a Manchester audience that forty hours before he had been in Brighton.[101] Such exertions took their toll. Having spoken on successive days in January 1839 in Edinburgh, Paisley and Glasgow, and after three hours sleep, he took a coach to Carlisle and spoke:

The large room was crowded to suffocation. We had good speeches. About 200 well-dressed females were present, who did me the honour to present me with a very beautiful scarf of their own manufacture, and tastily embroidered with their own hands ... The meeting went off triumphantly.

I went to bed very late, and rose at eight o'clock in order to reach Newcastle ... Upon getting up, I felt rather queer, and, upon sitting down to breakfast, I found that I had a violent pain in my chest, and no appetite, which with me is unusual. The pain in my chest increased and I felt a very disagreeable taste, upon which, I left the room and discovered that I had ruptured a blood vessel, either in the chest or upon the lungs. I was very sorry, for I did wish to see Universal Suffrage. I discharged about a wine glass of blood, and set off for Newcastle. Reached Newcastle at two, saw Dr Hume, who advised me by no means to speak, so I decided upon having a short sketch of my tour written and read to the meeting, but the visitors were so numerous, and the tidings so good, that eight, the hour of the meeting, had arrived, without narrative,

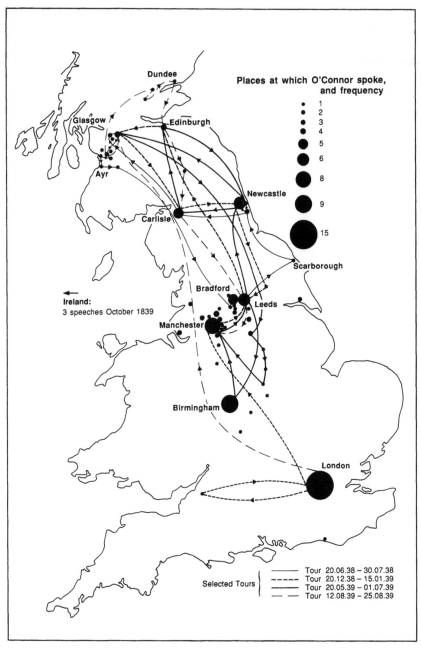

Places at which O'Connor spoke, and frequency

•	1
•	2
•	3
•	4
●	5
●	6
●	8
●	9
●	15

Ireland:
3 speeches October 1839

Selected Tours
———— Tour 20.06.38 – 30.07.38
- - - - Tour 20.12.38 – 15.01.39
———— Tour 20.05.39 – 01.07.39
— — Tour 12.08.39 – 25.08.39

5.4   Movements of Feargus O'Connor, 1838 and 1839
(*Source:* 'Northern Star')

and contrary to the advice of my doctor, I set off to the meeting, in the new and spacious Music Hall, which was overflowing. I spoke for more than an hour, and felt very weak, however, I got through and lived to hear the resolutions below passed unanimously.[102]

O'Connor was remarkably explicit about the reasons for his incessant travelling. He speaks more than once of the national movement as a chain that he is forging together through his travels:

[O'Connor] could not conclude without expressing his delight at having thus perfected the great chain between London and Edinburgh and Glasgow. All the links were now perfect. London, Newcastle, Carlisle, Glasgow, and Edinburgh had now become forged as it were together, and although the wages of corruption were taken from the provinces to support the idle in the metropolis, yet a spirit was now growing up which nothing but justice could put down.[103]

O'Connor's travelling and thus his personal acquaintance with radical leaders in all parts of the country, together with the circulation of his newspaper, the *Northern Star*, proved a necessary substitute for a formal national organisation.[104]

Besides such national figures there was a small army of agitators making tours such as those of Robert Gammage in 1842–3. Other examples abound: very similar to Gammage were Robert Lowery, a tailor caught up in an unplanned speaking tour of northern England in 1838 after going to London as a Chartist delegate from Newcastle,[105] and Henry Vincent, a Hull compositor who moved to London in 1833 and became a 'missionary' to Monmouthshire, the southwest, Banbury, and Hull;[106] both were paid by the London Working Men's Association. Others travelled mainly within a region, such as Richard Pilling, a central figure in the strikes of 1842, who in three weeks 'addressed upwards of 300,000 in different parts of Lancashire and Cheshire'.[107] One traveller described how 'the principles of Socialism…through Socialist lecturers, were spreading through the manufacturing districts'.[108] This propagation of radicalism paralleled that of temperance and religious movements, and some travellers were active in several fields.[109]

In earlier times these lesser missionaries were more obscure, but they included Joseph Mitchell, a Liverpool printer, and William Benbow, a Manchester shoemaker, despatched by a meeting held at Middleton in Lancashire to hold meetings through Yorkshire and the Midlands in December 1816. Mitchell 'moved frequently between London, the Midlands, and the North', visiting town after town, accompanied for some of his journeys by Oliver the spy.[110] Benbow spoke at Pentridge prior to the 1817 Rising, fled to New York in 1818 and returned in 1819 with Tom Paine's bones. He became a leader among London radicals and travelled again in the north after 1832.[111]

However, alongside these men, at least semi-professionals, were innumerable politically aware travellers, many of them artisans, whom we can only identify through autobiographies. The example of Robert Gammage has already been discussed at some length but several others can be given. William Adams was very active in Cheltenham Chartism in the 1840s, meeting Gammage when he stood for parliament there in 1852. When in 1855 he had to tramp home from the Lake District following the collapse of a republican magazine, he handed out tracts along the way.[112] John Buckmaster, having run away from Uxbridge to Tiverton, became active in both Chartism and the Anti-Corn Law League, becoming local secretary of the latter.[113] Like Gammage, James Hillocks had to leave home because of his involvement with the Chartists in Dundee,[114] while John Leno founded branches in both his home town of Uxbridge and in Windsor.[115] One radical shoemaker, first forced to leave London to escape arrest as one of the Cato Street conspirators, later moved from Northampton to Nottingham in 1830 and 'within a week of my arrival, I was what in the Fenian movement would be called a head centre'.[116] Still earlier, Thomas Preston's autobiography was written in 1817 following his acquittal on charges of high treason after the Spa Fields demonstrations in London the previous year, but prior to this 'a strong inclination for travelling' had taken him in the 1790s to the north of England and then to Dublin and Cork, where his leadership of a strike forced him to return to England.[117]

It is obviously hard to assess the relative impact of one famous orator such as O'Connor as opposed to that of many thousands of individual travellers such as Gammage or Adams, who took radical ideas wherever they went, but the latter were clearly significant. All these 'agitators' were travellers, swept up in the incessant geographical movement of the age, their horizons national, not local, because this was the arena within which their lives were played out. The emergence of large numbers of people in specialised occupations, forced to move extensively to find vacancies, together with the constant structural and cyclical changes in the economy, created these travellers and so made inevitable the emergence of national popular political movements.

Such levels of movement changed individuals in several ways. It gave them a network of social contacts from which broader movements could be constructed. The acquisition of such a spatially dispersed social network inevitably worked to break down the local chauvinism, even tribalism, that vitiated radical movements in less developed societies. The experience of travel itself would work to increase an individual's confidence in his ability to act effectively in varied and novel situations: finding oneself frequently in the uncomfortable situation of being surrounded by strangers, with whom one's relationship is undefined, is perhaps the best preparation for confrontations with other, and in conventional terms superior, social groups, where existing relationships must be redefined. Inevitably, it was the individuals who had been transformed in this way who would take the lead in the construction of new political movements; while the community might look to their oldest inhabitants to maintain their traditions, when change was sought the newcomer was often the leader. Further, it would be the organisations that already operated on the national scale, namely the artisan trade unions, that would provide the institutional support for such a national labour movement.

Mobility and class formation

The development of social history, labour history and historical geography over the last twenty years has given far greater emphasis to the study of communities, and this is reflected in several of the other essays in this volume. In many ways this has been a beneficial development: it has moved the emphasis away from institutions to individuals, from a simplistic economism to an involvement with many different strands of social theory and from narrow thematic studies to the broad span of human experience. None of this would I wish to see lost, but the location of all research within tightly bounded geographical areas inevitably excludes a vital part of individual experience: the experience of moving between areas. While this essay has been concerned with only one rather specific type of movement, limited to a particular period and class of person, it has suggested that this pattern of movement had significant consequences for the development of popular political movements in Britain. However, the study of the social consequences of geographical movement is one that can be developed in quite different historical contexts and is arguably an area where geographers can and should make a special contribution.

I am not arguing for a rejection of the scale and aims of the classic community study, although we can perhaps seek to explore ways of making the geographical focus slightly less rigid. However, if we wish to embrace crucial elements of individual experience, we must seek to extend the community study and, in particular, see migrants not as irritating individuals who keep dropping out of our linkages, or less formal assemblages of records, but as crucial agents both of change and of maintaining the community's links with a wider world. One work that organises a study of labour history around an individual is Prothero's account of John Gast and London artisan politics, but although Gast was born in Bristol and moved to Portsmouth after serving his apprenticeship, partly because of 'overbearing masters', Prothero's account is very much concerned with London industries and London politics.[118] Another starting point might be the careers of Robert Gammage and William Adams, both members of classic tramping crafts (coachmaking and printing), both active in the Chartist movement, both involved in the Cheltenham election in 1852, both spending much of their later life in Newcastle. By using the careers of such individuals we can begin to weave together craft and place, the local scale and the national, the experience of travelling and the importance of coming home.

This essay began with Eric Hobsbawm, and it is appropriate that it should end with Edward Thompson. Most discussion of the central thesis of *The making of the English working class* has focussed on whether there was a single working class, as distinct from working classes, but there is a parallel question: whether and how a specifically English, meaning national, working class was made, with solidarities and institutions overriding purely local loyalties. This question has been largely ignored because of the emphasis on local studies in subsequent research on both Thompson's period and the Chartists.[119] However, it is implicit in both Thompson's title and the range of his coverage; if he gives greater emphasis to Yorkshire than to the southwest, his geographical scope is far wider than that of most of his critics and his balancing of London and the north is exemplary.

Despite this, Thompson provides no direct answer to the question. Perry Anderson, in particular, makes this a central criticism and ties it to Thompson's failure to relate class formation to industrialisation and capital accumulation: 'there is no spatial map of British capitalism'.[120] One answer is provided by Langton, who argues that regional specialisation in production, central to the industrial revolution, served to emphasise the distinctiveness of localities and so fragmented working-class movements.[121] An alternative answer is that while the industrial revolution created some extremely specialised communities, notably the mill town and the mining village, it also created, as Thompson puts it, an artisan culture that had:

from Tudor times onwards ... grown more complex with each phase of technical and social change ... This was, perhaps, the most distinguished popular culture England has known. It contained the massive diversity of skills of the workers in metal, wood, textiles and ceramics, without those inherited 'mysteries' and superb ingenuity with primitive tools the inventions of the Industrial Revolution could scarcely have got further than the drawing board.[122]

Most of these highly differentiated skills were spatially dispersed if not ubiquitous and so, as this essay has argued, the reproduction of the workforce and the efficient operation of the labour market made necessary a high level of routine geographical mobility, embodied in cultural practices. This in turn brought forth an artisan sense of community that was national rather than local, and popular political movements with a national programme. Implicit in this account is a parallel set of local working classes, of the mills and the mines, but they, of course, are neither my subjects nor Thompson's.

Acknowledgements
The reading of autobiographies central to this essay was mainly done during a six-week stay as a Summer Scholar at St John's College, Oxford: I am grateful to the college for giving me the opportunity for such a sustained piece of reading. The analysis of the Annual Reports of the Steam Engine Makers' Society was a collaborative project with the Literary and Linguistic Computing Centre, Cambridge University; the Wellcome Trust for the History of Medicine funded the purchase of death certificates. I am grateful to Andrew Charlesworth, James Epstein, David Gilbert, Jack Langton, Robert Leeson, Robin Pearson, Adrian Randall, David Vincent and my collaborators for comments and suggestions.

Notes

Introduction: class, community and the processes of urbanisation

1 Law, 'The growth of urban population'.
2 There were, for example, twice as many Irish in New York in 1875 as in London in 1871, although London was three times as big as New York; and whereas only 9 per cent of the population of London had been born outside England and Wales, 58 per cent of the population of New York had been born outside the United States. Data on London from British Parliamentary Papers (BPP) 1873 [C. 872] 'Census of England and Wales. Vol. III'; and on New York City from *Census of the State of New York for 1875* (New York, 1877).
3 Carter, *Urban historical geography*; Vance, *This scene of man*.
4 Dennis and Prince, 'British urban historical geography'.
5 Warner, *Streetcar suburbs*; Thernstrom, *The other Bostonians*; Hershberg, *Philadelphia*; Zunz, *The changing face of inequality*.
6 Darby, 'The movement of population'; Lawton, 'Population trends'; Pooley, 'Welsh migration'.
7 Vance, 'The employment linkage'; Vance, 'Determinant and contingent ties'.
8 Gilbert, 'Pioneer maps'; Howe, *Man, environment and disease*.
9 McKenzie, 'The ecological approach'; Park and Burgess, *Introduction to the science of sociology*.
10 Wolff, *The sociology of Georg Simmel*; Park, 'The crowd and the public'; Reissman, *The urban process*; Canetti, *Crowds and power*; Sartre, *Critique of dialectical reason*.
11 See, for example, Max Weber's reactions to Alfred Weber's proposed study of workers' 'selection and adaptation' in heavy industry in Zohn, *Marianne Weber – Max Weber: a biography,* pp. 329–31; Hughes, *Consciousness and society*.
12 Schnore, 'Geography and human ecology'.
13 Hauser and Duncan, *Methods of urban analysis*; Duncan and Duncan, *The negro population of Chicago*; Shevky and Bell, *Social area analysis*.
14 Peach, *Urban social segregation*.
15 For British work, see Robson, *Urban analysis*; Pritchard, *Housing and spatial structure*; and the works discussed in Dennis, *English industrial cities*, ch. 1.
16 Robson, *Urban analysis*, p. xi.
17 Ward, *Poverty, ethnicity and the American city*; Harvey, *The condition of postmodernity*.

18 Dyos, *Camberwell*; Beresford, 'Prosperity Street'; Beresford, *East End, West End*; Dyos and Reeder, 'Slums and suburbs'.

19 Dennis, *English industrial cities*, ch. 5; Springett, 'Landowners and urban development'; Ward, 'The pre-urban cadastre'.

20 Hobsbawm, *Labouring men*; Gray, *The labour aristocracy*; Crossick, *An artisan elite*; Foster, *Class struggle and the industrial revolution*.

21 E. P. Thompson, *The making of the English working class*; id., 'Moral economy'.

22 Anderson, *Arguments*; Gregory, 'Contours of crisis'; McLennan, 'E. P. Thompson'.

23 Chevalier, *Labouring classes*; Briggs, 'Cholera and society'; Rosenberg, *The cholera years*; Morris, *Cholera 1832*; Durey, *The return of the plague*.

24 Briggs, *Victorian cities*; Hennock, *Fit and proper persons*; Meller, *Leisure*. For the argument that historical geography was left behind by these developments in urban history, see Fraser and Sutcliffe, 'Introduction' and Dennis and Prince, 'British urban historical geography'.

25 Whitehand, 'Building activity'; Daunton, 'The building cycle'; Whitehand, 'A reply'; Rodger, 'The building cycle'.

26 Ward, 'Victorian cities: how modern?'; Cannadine, 'Victorian cities: how different?'.

27 Cannadine, 'Residential differentiation'; in a collection of essays that reported a joint conference of urban historians and historical geographers, Johnson and Pooley, *The structure of nineteenth-century cities*.

28 See, for example, Billinge et al., *Recollections*.

29 Brookfield, *Interdependent development*; Harvey, *Social justice*.

30 See, for example, Gregory and Urry, *Social relations and spatial structures*; Held and Thompson, *Anthony Giddens*.

31 Gregory, *Ideology, science and human geography*.

32 Pooley, 'Book review'.

33 Berry, 'Cities as systems'.

34 Pred, *City-systems*; Robson, *Urban growth*; Chalklin, *The provincial towns*; de Vries, *European urbanisation*; Bairoch, *Cities and economic development*.

35 Although there have been interesting studies of the relations between towns and their hinterlands; Sharpless and Lindstrom, 'Urban growth'; Langton, 'Liverpool and its hinterland'.

36 Much the same point has been made for the early modern town by Langton and Hoppe, *Town and country*.

37 Population mobility is well attested by: Samuel, 'Comers and goers'; Wrigley, 'London's importance'; Clark and Souden, *Migration and society*.

38 See, for example, Langton and Hoppe, *Town and country*.

39 Consider, for example, the effect of industrial towns on rural wages in the north of England; Hunt, 'Wages'.

40 Williams, *The country and the city*; Webb, *From custom to capital*; Wiener, *English culture*; Thomas, *Man and the natural world*; MacFarlane, *The culture of capitalism*.

41 The best work by a historical geographer in this vein is Ward's superb studies of the social construction of the slum and the ghetto as phantoms in the middle-class mind, far more than just physical places on the ground: 'The Victorian

slum'; 'The ethnic ghetto'; *Poverty, ethnicity and the American city*; see also Driver, 'Moral geographies'.

42 These uses of the census enumerators' books have their place and they are well discussed in: Dennis, *English industrial cities*; Wrigley, *Nineteenth-century society*; Lawton, *The census and social structure*.

43 Recent work in geography along these lines includes: Thrift and Williams, *Class and space*.

44 Ward, 'Environs and neighbours'.

45 Ward, 'The ethnic ghetto'.

46 The reference to Charles Booth's celebration of a populace perpetually in movement and his abhorrence of areas where the residents seemed to stay put was suggested by a seminar given by David Reeder on Booth's image of London.

47 The literature of this topic is voluminous and may be sampled in the following: Calhoun, 'Community'; Sider, 'The ties that bind'; Anderson, *Imagined communities*.

48 See, for example, Pooley, 'The residential segregation', where the mere existence of Welsh-language chapels is used as evidence that there was a coherent and distinct Welsh culture in the city of Liverpool, whereas the relative scarcity of specifically Irish-language churches is taken to show the Irish people's failure to shape such a separate identity. It is, of course, a common mistake to use the existence of language as a marker for the existence of culture, a point elaborated upon in Chapter 3.

49 See, for example: Harrison, *Drink and the Victorians*; McLeod, *Class and religion*; Donajgrodski, *Social control*; Storch, 'The policeman as domestic missionary'.

50 The most exciting works on capitalist urbanisation have come from Harvey, *Consciousness and the urban experience*. The spatial restructuring of capitalism in relation to changing localities has been discussed in Massey, *Spatial divisions*; Rose, 'Locality-studies'.

1 Biology, class and the urban penalty

1 Kearns, 'Zivilis or Hygaeia'.

2 Kearns, 'Private property and public health reform', 'Cholera, nuisances and environmental management'.

3 Driver, 'Moral geographies'.

4 Cf. Flinn, *Report on the sanitary condition*, pp. 63–6. It is clear that contemporary French hygienists were closer to Alison's than to Chadwick's position: cf. Coleman, *Death is a social disease*; Delaporte, *Disease and civilisation*; la Berge, 'The French connection'.

5 Mayhew, *London labour II*, pp. 250, 248.

6 Mayhew, *London labour II*, pp. 223, 243, 249, 248.

7 Armstrong, 'The use of information about occupation'.

8 Crossick, *An artisan élite*. British Parliamentary Papers (BPP) 1904 [Cd. 2174] *Census of England and Wales, General Report with Appendices*.

9 Data for dates before 1881 exclude the dock and railway workers since they were only separately returned from that date.

10 There were only eleven registration divisions covering England and Wales. As urban, I have taken those with the lowest proportion of their male workforce in agriculture (Northwest, North, Yorkshire) and as rural those with the highest (East, South Midlands, Southwest). This degree of aggregation is forced on us by the inadequate occupation information in the published census tables. The Decennial Supplement gives data for a group of urban registration districts and for the registration divisions: *Supplement to the Thirty-Fifth Annual Report of the Registrar General 1871*, pp. 447–504.

11 From *Supplement to the Sixty-Fifth Annual Report of the Registrar General of Births, Deaths and Marriages in England and Wales. Part II*, pp. cxlii–clvii.

12 Kearns, *Urban epidemics*.

13 One further piece of evidence indirectly reinforcing these conclusions is provided by the fact that for each date we can study there was, as will be seen below, virtually no difference in age-specific mortality between farmers and farm labourers and, as Figure 1.2 shows, virtually no difference in their leading causes of death. Although these were rather heterogeneous groups, we might expect their average social status to differ somewhat.

14 Based on the Registrar General's Supplements for 1851–60 and 1891–1900. London is the registration division. The large towns are the registration districts covering the cities with a population of 50,000 or more in 1851. The rural set of districts are those sixty-four with more than 60 per cent of the male workforce in agriculture in 1851; the same districts were taken for 1891–1900, with the exception of Farnborough which had been absorbed into another district.

15 Based on *Supplement to the Sixty-Fifth Annual Report of the Registrar General of Births, Deaths and Marriages in England and Wales. Part II*, pp. cci–cciv and occupational tables in the censuses of 1861, 1871, 1881, 1891 and 1901.

16 BPP 1911 [Cd. 5077] 'Royal Commission on the Poor Laws and relief of distress. Appendix Volume XXV. Statistics relating to England and Wales'; in conjunction with the 1901 census.

17 Kearns, Lee and Rogers, 'The interaction of political and economic factors'.

18 BPP 1861 (474-II) ix, p. 731, 'Select Committee on the Poor Laws. Fifth report', pp. 40, 78.

19 See Fridlizius, 'Sex-differential mortality'.

20 Based on the published census volumes. The rural counties are the five with the highest proportion of males employed in agriculture in 1851 (Herefordshire, Huntingdonshire, Lincolnshire, Suffolk, Bedfordshire). The precise figures were – of the population aged sixty-five and over in 1851 in London, 59 per cent were female, 37 per cent widows, 14 per cent widowers and in rural counties 53 per cent, 29 per cent, 16 per cent respectively. In 1901 the corresponding figures for London were 60 per cent, 37 per cent, 14 per cent and for the same rural counties 54 per cent, 29 per cent, 15 per cent.

21 Data came from BPP 1910 [Cd. 4983] 'Royal Commission on the Poor Laws and relief of distress. Appendix Volumes XII. Memoranda by individual commissioners on various subjects', pp. 366–7; BPP 1911 [Cd. 5077] 'Royal Commission on the Poor Laws and relief of distress. Appendix Volume XXV. Statistics relating to England and Wales', Tables 4 and 7.

22 Lampard, 'The history of cities', 'The urbanising world', 'The nature of urbanisation'.

23 Chevalier, *Labouring classes*, p. 5.
24 Kearns, 'Death in the time of cholera'.
25 Coleman, *The idea of the city*.
26 Charles Dickens, *Bleak house* (1853. Harmondsworth: Penguin, 1971), p. 705.
27 Charles Dickens, *Dombey and son* (1848. Oxford: Oxford University Press, 1974), pp. 402, 692, 625, 541.
28 Williams, *The rich man and the diseased poor*.
29 Flinn, *Report on the sanitary condition*, pp. 219, 223, 422, 78.
30 Ibid., p. 423.
31 Mayhew, *London labour I*, pp. 2–3.
32 See also: Hamlin, 'Providence and putrefaction'.
33 Mayhew, *London labour II*, pp. 160, 217, 464, 394, 423.
34 Williams, *The rich man and the diseased poor*, p. 111.
35 *Dombey and son*, pp. 93, 191, 236, 404–5, 37, 365.
36 *Bleak house*, pp. 683, 547–8.
37 *Dombey and son*, p. 356; *Bleak house*, p. 271.
38 *Bleak house*, pp. 273, 50.
39 Schwarzbach, *Dickens and the city*, p. 153.
40 Charles Dickens, *Our mutual friend* (1865. Oxford: Oxford University Press), p. 21).
41 Quoted in Schwarzbach, *Dickens and the city*, p. 211.
42 Health of Towns Association, *Address from the Committee of the Health of Towns Association to the Right Reverend the Bishop of London and the reverend clergy* (London: Health of Towns Association, 1848), p. 4.
43 Quoted in Finer, *Chadwick*, p. 222.
44 See Heath, *The sexual fix*.
45 'Il est très clair qu'une physiologie sociale de l'excrétion constitue le thème directeur de la pensée de Parent-Duchâtelet. L'excrétion fécale recueillie par les égouts est indispensable au fonctionnement de la ville, comme l'excrétion séminale que recueille la prostituée.' Lécuyer, 'L'hygiène en France avant Pasteur', p. 126.
46 Gay, *The education of the senses*, p. 317.
47 Stallybrass and White, *The politics and poetics of transgression*, pp. 191, 200.

2 Public space and local communities: the example of Birmingham, 1840–1880

1 This architectural tradition is discussed in Swenarton, 'Notes on the concept'.
2 This approach by historical geographers is to be seen in: Pooley, 'The residential segregation'; Shaw, 'The ecology'.
3 The literature that gives prominence to circulation, accumulation and social reproduction in cities includes: Saunders, *Social theory*; Badcock, *Unfairly structured cities*.
4 Similar observations in relation to the implications of residential segregation on class relations in nineteenth-century towns are made in: Cannadine, 'Residential differentiation'; Harris, 'Residential segregation'.
5 A point made in Jones, 'Class expression'.
6 Such as Briggs, *Victorian cities*; Reid, 'Interpreting the festival'.

7 F. M. L. Thompson, 'Introduction'; Olsen, *The growth*, pp. 23–4, 221–2, 236; Davidoff and Hall, 'The architecture'.
8 White, *Rothschild buildings*; id., *The worst street*.
9 Ross, 'Survival networks'; Humphries, *Hooligans or rebels?*; Woods, 'Community violence'.
10 Daunton, 'Public place', pp. 212–33, 218.
11 Ibid., p. 219.
12 See Daunton, *House and home*.
13 Meacham, *A life apart*, pp. 44–52.
14 Roberts, *A woman's place*, pp. 169–201.
15 Bushaway, *By rite*.
16 Poole, 'Oldham wakes'; Walton, 'The Lancashire wakes'.
17 Cunningham, 'The metropolitan fairs'; Reid, 'Interpreting the festival'.
18 Hammerton and Cannadine, 'Conflict and consensus'; Parker, 'The changing character'.
19 Harrison, 'The ordering'.
20 Jones, *Crime, protest*.
21 Bailey, 'Will the real'; id., *Leisure and class*.
22 Malchow, 'Public gardens'.
23 Daunton, 'Public place', p. 214.
24 Ibid., pp. 212–33.
25 Bunce, *History of the corporation of Birmingham*, II, pp. 76–8; *Birmingham Morning News*, 11 January 1875, p. 3.
26 Based on systematic samples for the borough of Birmingham, but excluding suburban Edgbaston, of one in fifteen household heads in 1851 and one in twenty household heads in 1871 and including for each of these cases the preceding household head in the same street or court. The census enumerators' books used in this analysis were HO 107 2051 to HO 107 2061 and RG 10 3089 to RG 10 3153. *Birmingham Morning News*, 4 February 1885, p. 5; Woods, 'Mortality and sanitary', pp. 53–5.
27 White, *Life in a court*, p. 3.
28 Brindley, *Church work*, pp. 7–9; *Birmingham Daily Press*, 22 June 1855, p. 2.
29 Matthison, *Less paint*, p. 65.
30 Special meeting of the Council to consider the report of the Sewerage Inquiry Committee, *Birmingham Council Proceedings*, 26 October 1871; *Aris's Birmingham Gazette*, 24 December 1849.
31 White, *Life in a court*, p. 2.
32 *Birmingham Daily Press*, 13 January 1857, p. 3. It seems in some cases, however, arrangements were made between tenants to use the washhouse on different days – see Church of the Messiah, *Ministry to the poor report for 1875–6*, p. 44.
33 Newman, *Defensible space*, pp. 3–66.
34 *Aris's Birmingham Gazette*, 24 December 1849.
35 Brindley, *Church work*, p. 79.
36 *Birmingham Daily Mail*, 5 June 1871, p. 4; *Aris's Birmingham Gazette*, 24 December 1849; '"I remember" reminiscences of Birmingham's public men', *Cuttings from the Birmingham Gazette and Express*, 1907–8, p. 136; Wilkinson, *Rough roads*, p. 2.
37 Daunton, 'Public place', pp. 218–19.

38 *Birmingham Daily Mail*, 11 May 1871, p. 3.
39 Ibid., 3 April 1871, p. 4.
40 Ibid., 5 June 1871, p. 4; *Aris's Birmingham Gazette*, 24 December 1849.
41 Similar arguments are advanced in McLeod, 'White collar', in Ross, 'Survival networks' and in Roberts, *A woman's place*, pp. 183–201.
42 *Birmingham Morning News*, 26 March 1874, p. 7.
43 *Birmingham Daily Mail*, 23 July 1873, p. 3.
44 Ibid., 19 October 1871, p. 4, and 26 October 1871, p. 4.
45 For example, *Birmingham Mercury*, supplement, 29 August 1857, p. 2; *Birmingham Journal*, 26 May 1855, p. 6, and 2 June 1855, p. 6.
46 *Birmingham Mercury*, 4 January 1851.
47 Ibid.
48 For a discussion of the social importance of public ritual, see Turner, *The ritual process*.
49 *Birmingham Daily Mail*, 19 October 1871, p. 4, and 26 October 1871, p. 4.
50 *Birmingham Journal*, supplement, 5 July 1856, p. 4.
51 *Birmingham Morning News*, 19 April 1875, p. 4.
52 Ibid., 22 July 1875, p. 6.
53 Lloyd, *Sketch*, p. 26.
54 Ibid., p. 18; *Report of the Artisans' Dwellings Inquiry Committee for presentation to the Council at its meeting on 3 June 1884*, p. 14.
55 British Parliamentary Papers (BPP) 1865 [3548] xx, p. 103, 'Fourth report of the Children's Employment Commission', p. 225.
56 Brindley, *Church work*, p. 118.
57 The Irish in Birmingham formed a relatively small proportion of the population (in 1851 only 5.5 per cent of Birmingham's adults had been born in Ireland) and their residential concentrations seem to have been small. BPP 1852–3, lxxxv, '1851 census of population', p. 526. For examples of such 'Irish rows', see *Birmingham Journal*, 16 March 1850, p. 3; *Birmingham Daily Gazette*, 19 June 1862, p. 2, 7 July 1862, p. 2, and 5 September 1862, p. 2. For examples of names given to Irish enclaves, see *Birmingham Journal*, 12 March 1856, p. 3, and 3 May 1856, p. 7.
58 *Birmingham Daily Mail*, 13 June 1871, p. 2.
59 For a discussion of the problems of interpretation resulting from the geographical interspersal of Irish districts with disorderly pubs and brothels, in this case in York, see Finnegan, *Poverty and prejudice*, ch. 9.
60 For a general discussion of police powers in several towns in the nineteenth century, see Storch, 'The policeman'.
61 Accounts of the organisation of Birmingham's police include Stephens, *The city of Birmingham*, p. 337; Bunce, *History of the corporation of Birmingham*, vol. I, pp. 81, 184–271, 281–6, and vol. II, pp. 254–99; Gill, *History of Birmingham*, pp. 274–7; Stephenson, thesis, pp. 178–9. The actual numbers of law enforcers may have been slightly less than the authorised strength of the police force. The 1857 and 1876 populations were derived by assuming equal annual increases between relevant decennial censuses.
62 A point made in a more general context by Woods, 'Community violence', pp. 180–1, 183.
63 Thorne, *My life's battles*, pp. 26–7.

64 *Birmingham Daily Mail*, 15 May 1871.

65 Examples of such 'rescues' include BPP 1877 [171] xi, p. 1, 'Report from the Select Committee of the House of Lords on prevalent habits of intemperance. First report', Q. 2133; *Birmingham Daily Mail*, 23 July 1873, p. 3; *Birmingham Morning News*, 12 January 1875, p. 8; *Birmingham Daily Post*, 8 March 1875; *Birmingham Daily Gazette*, 1 March 1875, p. 6, and 10 March 1875, p. 4; *Birmingham Journal*, 6 April 1850, p. 3, *Birmingham Morning News*, 26 October 1875, p. 4.

66 Bunce, *History of the corporation of Birmingham*, vol. II, pp. 294, 302; *Report of the police establishment and the state of crime in the borough of Birmingham for the year ending 29 September 1878*, pp. 7–8. The police numbers are based on the authorised strength of the police force – actual numbers may have been slightly less. The 1877 population was derived by assuming equal annual increments between the 1861 and 1871 censuses.

67 A similar case is made by Weinberger, although she argues for an upward trend in police assaults in Birmingham between 1863 and the early 1870s. She tends to unfairly exaggerate this trend, however, by using only the absolute number of assaults. When these assaults are expressed as a ratio of the town's increasing population, the rise is rather less clear. Weinberger, thesis, pp. 195–6, and 'The police', pp. 67–8, 86–9.

68 Ibid., pp. 65–93. Occupational information was extracted from daily coverage of the Birmingham Police Court in *Aris's Birmingham Gazette* for October 1867 to October 1868 and October 1876 to October 1877, and this is compared to a 1 per cent sample of the population aged over 10 in the 1871 census for Birmingham.

69 Weinberger ('The police', pp. 67, 89) makes the same proposal, again for Birmingham.

70 *Birmingham Daily Post*, 2 April 1863, p. 3.

71 *The Town Crier*, April 1870; Weinberger, thesis, p. 96.

72 *Birmingham Daily Mail*, 23 June 1873, p. 3, and 20 December 1873, p. 2.

73 Reid, 'Interpreting the festival'.

74 *Birmingham Daily Mail*, 23 November 1871, p. 4, and 25 May 1871, p. 4; Jaffray, *Hints for a history of Birmingham*, ch. 26. Examples of the remnants of prize fighting in Birmingham include *Birmingham Daily Mail*, 9 April 1872, p. 3, and 19 November 1873, p. 3; *Birmingham Journal*, 6 September 1851, p. 5.

75 *Birmingham Daily Mail*, 20 July 1871, p. 3, and 9 May 1871, p. 2; *Birmingham Daily Post*, 18 September 1863, p. 2.

76 *Birmingham Daily Mail*, 23 November 1871, p. 4.

77 *Birmingham Mercury*, 4 January 1851.

78 A similar observation is made on the public censure of individuals in the workplaces in Birmingham by Behagg, 'Secrecy, ritual and folk violence', p. 165.

79 For a discussion of the activities associated with occasional more sustained drinking, see Martin, *A sociology*, pp. 71–3. For examples of 'Irish rows' in some streets, see Finigan, *Journal of the Birmingham Town Mission missionary*, p. 79; *Birmingham Mercury*, 3 May 1856, p. 3; *Birmingham Journal*, 30 August 1851, p. 5; *Birmingham Daily Post*, 10 August 1860, p. 2, and 28 September 1863, p. 3; *Birmingham Morning News*, 15 July 1874, p. 3; *Birmingham Daily Mail*, 13 June 1871, p. 2.

80 Daunton, 'Public place', p. 223.
81 BPP 1873 [C. 872], p. 1, 'Census of England and Wales 1871. Population abstracts. Vol. III', p. 342, table 21.
82 For instance, New Meeting Ministry to the Poor, *Annual report of the Minister for 1853*, p. 18, and Palfry, *Mission work*, p. 4.
83 From a description of Birmingham life in De Rousier, *The labour question*, p. 285; *Birmingham Daily Press*, 16 May 1855, p. 2.
84 Wilkinson, *Rough roads*, p. 28.
85 Thorne, *My life's battles*, pp. 26–7.
86 For instance, see *Birmingham Mercury*, supplement, 29 August 1857, p. 2; *Birmingham Daily Post*, 5 June 1860, p. 3, and 20 June 1860, p. 3; Palfry, *Mission work*, pp. 3–6; New Meeting Ministry to the Poor, *Annual report for 1848*, p. 31; *Birmingham Daily Mail*, 3 May 1871, p. 3; *Birmingham Morning News*, 11 February 1875, p. 5.
87 *Aris's Birmingham Gazette*, 24 December 1849.
88 This trend is also noted in Weinberger's study of crime in Birmingham between 1867 and 1877; see Weinberger, thesis, pp. 174–6, 196, 216–26, 230.
89 Adolescents in the nineteenth century are discussed in Walvin, *A child's world*, pp. 79–100, 149–58.
90 Matthison, *Less paint*, p. 63. Detailed evidence on the fashions of the street gangs in the period is contained in Pearson, *Hooligan*, pp. 92–101.
91 Territorial conflicts between street gangs in the period up to 1939 are examined in Humphries, *Hooligans or rebels?*, pp. 174–208. Working-class attitudes to violence in the nineteenth century are examined in Woods, 'Community violence', p. 171.
92 *Birmingham Morning News*, 18 February 1874, p. 3; Wilkinson, *Rough roads*, p. 28.
93 Matthison, *Less paint*, p. 63.
94 BPP 1870 [91] liv, p. 1, 'Schools for the poorer classes in Birmingham, Leeds, Liverpool and Manchester. Special reports', p. 74. On the Birmingham Education Society which undertook this census, see Hennock, *Fit and proper persons*, pp. 81–6.
95 Briggs, *A history*, vol. II, p. 106; Smith, *Conflict and compromise*. More generally, see Gardner, *The lost elementary schools*.
96 Stephens, *The city of Birmingham*, p. 493; Briggs, *A history*, vol. II, p. 106; *Birmingham School Board report on compulsion, as applied to school attendance in Birmingham* (1878), pp. 7–8; *Report of the work accomplished by the Birmingham School Board during the six years ended 28 November 1876*, p. 15; Birmingham School Board, *Annual report*, 1877, p. 11, and 1880, p. 11.
97 *Birmingham Morning News*, 30 January 1875, p. 6; *Birmingham School Board report on compulsion, as applied to school attendance in Birmingham* (1878), p. 9; *Birmingham Daily Gazette*, 15 September 1887; Smith, *Conflict and compromise*, p. 211.
98 *Birmingham School Board report on compulsion, as applied to school attendance in Birmingham* (1878), pp. 19–21.
99 Ibid., p. 43, table XI.
100 Ibid, pp. 13, 19, 22.
101 Smith, *Conflict and compromise*, p. 211; *Birmingham Daily Gazette*, 15 September

1887; *Birmingham School Board report on compulsion, as applied to school attendance in Birmingham* (1878), pp. 16–17, 22–8.

102 Ibid., p. 43, table XI, p. 22.
103 Daunton, 'Public place', pp. 212–33.
104 Meller, *Leisure*, pp. 97–124; Gaskell, 'Gardens'.
105 Douglas Reid has also studied the provision of municipal parks in Birmingham and has argued convincingly that the general emphasis of this public dialogue changed over the nineteenth century. He contends this provision was more likely to be advocated as a means of direct class conciliation in the 1840s and 1850s than in the 1870s, by which time the more elaborate and diffuse notions of civic advancement and of common citizenship were likely to be used, influenced by a receding prospect of working-class political unrest and by the 'civic gospel' in Birmingham in the 1870s. See Reid, thesis, pp. 285–322, 436, 465.
106 *Birmingham Daily Press*, 14 January 1856, p. 2.
107 Birmingham Town Council, *Baths and Parks Committee Minutes* (hereafter *BPCM*), 16 September 1861; *Birmingham Daily Press*, 14 January 1856, p. 2; *Birmingham Mercury*, 27 June 1857, p. 5.
108 *Birmingham Daily Press*, 16 May 1855, p. 2; *BPCM*, 7 February 1877.
109 *BPCM*, 18 September 1861.
110 Ibid., 6 December 1876.
111 Ibid., 6 May 1876, and 27 December 1859; *Birmingham Morning News*, 6 May 1875, p. 4.
112 *Birmingham Morning News*, 21 January 1875, p. 1.
113 Both Aston Park and Adderley Park had been parks during the late 1850s and early 1860s before they became municipal parks. For accounts of the chronology of Birmingham's parks, see Dent, *The making of Birmingham*, pp. 486–7; *Birmingham Daily Press*, 14 January 1856, p. 2, and 1 September 1856, p. 3; *BPCM*, 23 June 1862, 2 February 1863, 30 March 1863, 4 August 1863, 27 October 1863, 15 February 1864; Bunce, *History of the corporation of Birmingham*, vol. II, pp. 195–204.
114 Bunce, *History of the corporation of Birmingham*, vol. II, p. 204; *BPCM*, 7 February 1877.
115 A rare description of their developing geography – in another town – is provided in Strachan and Bowler, 'The development'. Among the best accounts of the unequal geography of public service provision in contemporary cities are Pinch, *Cities and services*; Kirby and Pinch, 'Territorial justice'; and Knox, 'Residential structure'.
116 *BPCM*, 4 February 1862, 5 May 1874, 4 August 1874, 9 November 1874, 9 February 1875. The population for 1874 was derived by assuming an equal annual population increase between the 1871 and 1881 censuses.
117 *BPCM*, 18 May 1858, and 30 May 1859.
118 Molyneux, thesis, pp. 26–7.
119 *Birmingham Daily Post*, 15 September 1863, p. 3.
120 *Birmingham Morning News*, 1 June 1875, p. 8; *Birmingham Mercury*, 26 July 1856, p. 5.
121 *Birmingham Daily Post*, 8 April 1863, p. 7.
122 *Birmingham Morning News*, 5 June 1874, p. 4, and 4 June 1874, p. 4.

123 BPP 1867–8 [402] xiv, p. 1, 'Select Committee on the sale of liquors on Sunday Bill', qq. 7991, 7994–5.
124 For descriptions of Sunday walks and visits to suburban pubs, see *Birmingham Daily Mail*, 18 April 1872, p. 4. On suburban drinking on Sunday, see *Birmingham Daily Mail*, 9 June 1873, p. 2, 12 June 1873, p. 4 and 17 June 1873, p. 4; Dent, *The making of Birmingham*, pp. 419–20. Details of dances held in some suburban pubs are given in *Birmingham Daily Mail*, 11 September 1872, p. 4.
125 'Sale of liquors on Sunday' (see note 123), q. 1749.
126 The view that working-class behaviour often adapted to the role expected in the specific social context without this becoming consistent subsequently is examined in Bailey, *Leisure and class*, pp. 176–80, and 'Will the real'.
127 The 'civic gospel' of the 1870s, embracing this rhetoric of 'civic community', is discussed in Hennock, *Fit and proper persons*; Jones, 'The public pursuit'.
128 Daunton, 'Public place', pp. 212–13; 218–19.
129 Meacham, *A life apart*, pp. 44–52.
130 Reid, 'Interpreting the festival'.
131 A similar argument is made in Urry, 'Society, space', and Calhoun, 'Class, place'.

3 Class, culture and migrant identity: Gaelic Highlanders in urban Scotland

1 See, for example, Lees, *Exiles of Erin*; Swift and Gilley, *The Irish in the Victorian city*; Lees, 'Patterns of lower-class life'; Gilley, 'English Catholic charity'; E. Jones, 'The Welsh in London in the nineteenth century'; Pooley, 'The residential segregation'; Pryce, 'Migration and the evolution of culture areas'; see also the references in Pooley, 'Welsh migration to England'; Jenkins Williams, *The Welsh church and Welsh people in Liverpool*; B. Williams, *The making of Manchester Jewry*; id., 'The beginnings of Jewish trade unionism in Manchester'; Werly, 'The Irish in Manchester'; Buckman, 'Alien working-class response'; Dillon, 'The Irish in Leeds'; Connell, 'The gilded ghetto'; Large, 'The Irish in Bristol'; Finnegan, 'The Irish in York'; Richardson, 'Irish settlement in mid-nineteenth century Bradford'; Cowlard, 'The urban development of Wakefield'; Roy Lewis, 'The Irish in Cardiff': see also Carter and Wheatley, 'Some aspects of the spatial structure'; Dennis, 'Community and interaction in a Victorian city: Huddersfield 1850–1880', pp. 134–7; id., *English industrial cities*, pp. 35–8, 225–6; Lobban, 'The Irish community in Greenock'; id., 'The migration of Highlanders into Lowland Scotland'; Collins, 'The social experiences of Irish migrants in Dundee and Paisley' (cf. Dennis, *English industrial cities*, p. 39); Withers, *Highland communities in Dundee and Perth*; Phillips, *A history of the first Jewish community in Scotland*; Aspinwall and McCaffrey, 'A comparative view of the Irish in Edinburgh'.
2 Lees, 'Patterns of lower-class life', p. 360.
3 British Parliamentary Papers (BPP) 1836 [40], xxiv, p. 427, 'Report on the state of the Irish poor in Great Britain'; J. G. Williams, 'The impact of the Irish on British labour markets'; Swift and Gilley, *The Irish in the Victorian city*, pp. 1–12.

4 Werly, 'The Irish in Manchester', p. 350; see also Gilley, 'The Catholic faith of the Irish slums'; Lees, *Exiles of Erin*, ch. 7.
5 'Report on the state of the Irish poor' (see note 3), pp. 436–68.
6 Dillon, 'The Irish in Leeds', pp. 17–18; Jackson, *The Irish in Britain*, pp. 46–8.
7 Pooley, 'Welsh migration to England', p. 298.
8 E. Jones, 'The Welsh in London in the nineteenth century', *passim*; see also his earlier work 'The Welsh in London in the seventeenth and eighteenth centuries'.
9 B. L. Williams, *The making of Manchester Jewry*, pp. 17–18, 24–5; Phillips, *A history of the origins*, pp. 6–7; Connell, 'The gilded ghetto', pp. 52–4; Mendelman, *The Jews of Georgian England*; Roth, *The rise of provincial Jewry*.
10 Fife and Power, *Black settlers in Britain 1585–1958*. On more general perspectives on other migrant groups not mentioned, see for example Holmes, *Immigrants and minorities in British society*, especially chapters by H. Kellenbenz, 'German immigrants in England' (pp. 63–80), and J. P. May, 'The Chinese in Britain' (pp. 111–24); Holmes, 'The German gypsy question in Britain, 1904–1906'; Lunn, 'Reaction to Lithuanian and Polish immigrants in the Lanarkshire coalfield, 1880–1914', in id., *Hosts, immigrants and minorities*, pp. 308–42; Hepburn, *Minorities in history*; Feldman, 'There was an Englishman, an Irishman and a Jew...: Immigrants and minorities in Britain'.
11 Swift and Gilley, 'Introduction' in id., *The Irish in the Victorian city*, p. 9; O'Tuathaigh, 'The Irish in nineteenth-century Britain'; Gilley, 'English attitudes to the Irish in England 1789–1900'. On the wider perspectives of racism and immigrant populations, see for example, Thurley, 'Satan and Sambo'; Kirk, 'Ethnicity, class and popular toryism, 1850–1870'; Lee, 'Aspects of the working-class response to the Jews in Britain, 1880–1914'. On a particular town and the forms taken in protest against the Irish there, see Millward, 'The Stockport riots of 1852: a study of anti-Catholic and anti-Irish sentiment'.
12 Holmes, 'J. A. Hobson and the Jews' in id., *Immigrants and minorities*, pp. 125–57; Buckman, 'Alien working-class response; Lee, 'Aspects of the working-class response'; May, 'The Chinese in Britain 1860–1914'.
13 Lees, 'Patterns of lower-class life', p. 370; B. Williams, *The making of Manchester Jewry*, pp. 22–3; Lees, *Exiles of Erin*, pp. 55–87.
14 This point is made in regard to migrant populations today by Feldman's (1983) historiographical review (see n. 10).
15 Pooley, 'The residential segregation', p. 367.
16 O'Tuathaigh, 'The Irish in nineteenth-century Britain', pp. 23–4.
17 Cosgrove, 'Towards a radical cultural geography'; J. S. Duncan, 'The superorganic in American cultural geography'; Withers, *Gaelic Scotland*, pp. 16–43.
18 Park, 'Human migration and the marginal man'. On urban culture and the idea of the 'marginal man', see also Castells, *The urban question: a Marxist approach*, pp. 75–85, 96–112; id. *City, class and power*, pp. 23, 34, 38–9; M. P. Smith, *The city and social theory*, pp. 177–80; Saunders, *Social theory and the urban question*, pp. 80–109.
19 Conzen, 'Immigrants, immigrant neighbourhoods and ethnic identity' for a very good review of historical immigrant identity in the United States (which concentrates on issues of residential pattern). On more modern formulations of urban ethnicity, see Yancey et al., 'Emergent ethnicity: a review and reformation'; Maldonado and Moore, *Urban ethnicity in the United States*.

20 Ward, 'The ethnic ghetto in the United States: past and present'; id., *Cities and immigrants*; id., *Poverty, ethnicity, and the American city, 1840–1925*.
21 J. S. Duncan, 'The superorganic in American cultural geography', p. 197; see also L. White, 'The concept of culture', p. 235; Schneider, 'The idea of culture in the social sciences' in Schneider and Bonjean, *The idea of culture in the social sciences*, p. 129.
22 R. Williams, 'Base and superstructure in Marxist cultural theory', pp. 6–7; id., *Marxism and literature*, p. 82; Cosgrove, 'Towards a radical cultural geography', pp. 5–6; Blaut, 'A radical critique of cultural geography', p. 27.
23 M. P. Smith, *The city and social theory*, pp. 219–20.
24 I have based my analysis of hegemony on the following: G. A. Williams, 'The concept of *egemonia* in the thought of A. Gramsci'; Bates, 'Gramsci and the theory of hegemony'; Femia, 'Hegemony and consciousness in the thought of Antonio Gramsci'; Adamson, *Hegemony and revolution*; Forgacs and Nowell-Smith, *Antonio Gramsci: selections from cultural writings*; R. Q. Gray, 'Bourgeois hegemony in Victorian Britain'; Gramsci, *Prison notebooks*; Hoare and Nowell-Smith, *Selections from the prison notebooks of Antonio Gramsci*; Buci-Glucksmann, *Gramsci and the state*; Cheal, 'Hegemony, ideology and contradictory consciousness'; Mouffé, *Gramsci and Marxist theory*; Nield and Seed, 'Waiting for Gramsci'; Showstack Sassoon, *Gramsci's politics*.
25 Williams, *Marxism and literature*, p. 110.
26 On this point, see Dupré, *Marx's social critique of culture*, pp. 216–75; P. Anderson, 'The antinomies of Antonio Gramsci', p. 22.
27 Neale, 'Cultural materialism', p. 209.
28 Dupré, *Marx's social critique of culture*, pp. 227–8.
29 I have elsewhere touched upon this point: Withers, *Gaelic Scotland*, pp. 332–7; see also Inglehart and Woodward, 'Language conflicts and political community'; Edwards, *Language, society and identity*; Neale, 'Cultural materialism', pp. 208–10; R. Williams, *Marxism and literature*, pp. 22, 34.
30 R. Williams, 'Base and superstructure', p. 8.
31 R. Williams, *Marxism and literature*, pp. 112–13.
32 E. P. Thompson, *The poverty of theory*, p. 74; id., 'Patrician society, plebian culture', pp. 388–9.
33 Billinge, 'Hegemony, class and power in later Georgian and early Victorian England', pp. 34–5.
34 Poulantzas, 'Marxist political theory in Great Britain'; Gray, 'Bourgeois hegemony in Victorian Britain'; Billinge, 'Hegemony, class and power', pp. 35–6 (original emphasis).
35 Mouffé, *Gramsci and Marxist theory*, pp. 179, 181, 193.
36 Agnew et al., *The city in cultural context*, pp. 7–8.
37 Ibid., p. 7.
38 Corrigan and Sayer, *The great arch: English state formation as cultural revolution*, pp. 6–7; see also Dentith, 'Political economy, fiction and the language of practical ideology in nineteenth-century England', pp. 183–99; Smail, 'New language for labour and capital'.
39 Jackson Lears, 'The concept of cultural hegemony: problems and possibilities', p. 568.
40 These paragraphs have been drawn from Devine, 'Highland migration to

Lowland Scotland, 1760–1860'; id., 'Temporary migration and the Scottish Highlands in the nineteenth century'; id., *The Great Highland famine*, chs. 6 and 8; Withers, 'Highland migration to Dundee, Perth and Stirling, 1753–1891'; id., 'Destitution and migration: labour mobility and relief from famine in Highland Scotland 1836–1850'.

41 Withers, 'Kirk, club and culture change: Gaelic chapels, Highland societies and the urban Gaelic subculture in eighteenth-century Scotland'; id., *Gaelic in Scotland 1698–1981: the geographical history of a language*, ch. 9.

42 BPP 1837–8 [109] xxxii, p. 1 'Second report of the Commissioners of Religious Instruction, Scotland', p. 459. The detailed evidence in this Report appears in the Appendices, although brief mention is made of the Gaelic chapels in Glasgow in the text of the Report. The material for Duke Street appears on pp. 32–6; for St Columba on pp. 60–4; and for West Gaelic Church on pp. 168–72 of Appendix II; in Appendix III, the pages covered for the three chapels are pp. 101–7, 179–98, 529–33 respectively. This evidence is referenced as 'Report of the Commissioners', with the Appendix and page number(s) following.

43 'Report of the Commissioners', Appendix III, p. 503.

44 Withers, 'Gaelic speaking in urban Lowland Scotland: the evidence of the 1891 Census'; id., 'Historical urban geolinguistics: Gaelic speaking in urban Lowland Scotland in 1891'.

45 Withers, 'Gaelic speaking', *passim*.

46 Edwards, *Language, society and identity*; Withers, *Gaelic in Scotland*, p. 107; Williams, *Language in geographic context*.

47 Withers, 'Highland clubs and Gaelic chapels'.

48 'Report of the Commissioners' (see note 42), Appendix III, p. 182.

49 BPP 1844 [557] xx, p. 1, 'Report from Her Majesty's Commissioners for inquiring into the administration and practical operation of the Poor Laws in Scotland, vol. I', p. xlix.

50 On this point see BPP 1841 [182] vi, p. 1, 'Report from the Select Committee appointed to inquire into the condition of the population of the islands and Highlands of Scotland, and into the practicability of affording the people relief by means of emigration', p. 47.

51 'Report of the Commissioners' (see note 42), Appendix III, p. 507.

52 Ibid., Appendix II, p. 189.

53 Brown, 'The costs of pew-renting'; see also Inglis, *Churches and the working classes in Victorian England*, pp. 48–57, 96–7, 105–8, 129–30; Gilbert, *Religion and society in industrial England*, pp. 69–93; McLeod, *Class and religion in the late Victorian city*, pp. 1–22.

54 'Report of the Commissioners' (see note 42), Appendix II, p. 119.

55 Ibid., Appendix III, p. 421.

56 Ibid.

57 Ibid., Appendix III, p. 505.

58 'Report from the Select Committee' (see note 50), pp. 118, 119.

59 Ibid., p. 118.

60 Withers, 'On the geography and social history of Gaelic', p. 125.

61 'The directory of Highland and Celtic societies'.

62 List of members, constitution and rules of the Uist and Barra Association [*Comunn Uidhist agus Bharraidh*], Session 1897–8, pp. 7–8. I am grateful to Mrs I. McCoard, current secretary of the Uist and Barra Association, for her permission to consult the records of this body.
63 Devine, *The Great Highland famine*, pp. 197 and 210.
64 From the General Applications for Parochial Relief, City Parish and Barony Parish, Glasgow (1855–91). These sources allow detailed insight into occupation and residences within the city as well as place of birth of Glasgow's poor Highland-born (and other pauper) population; Withers, 'Poor relief in Scotland and the General Register of Poor'. For general discussions of poor relief in nineteenth-century Glasgow, see Cage, 'The nature and extent of poor relief' in id., *Working-class in Glasgow*, pp. 77–97; Gourvish, 'The cost of living in Glasgow in the early nineteenth century'; Cage, 'The standard of living debate: Glasgow, 1800–1850'.
65 J. Smith, 'Class, skill and sectarianism in Glasgow and Liverpool, 1880–1914'.
66 Withers, *Highland communities in Dundee and Perth*, pp. 35–6.
67 Based on analysis of the census enumerators' schedules for Barony parish, 1851.
68 Dennis, *English industrial cities*, p. 273.
69 'Report of the Commissioners' (see note 42), Appendix II, p. 424.
70 Ibid., p. 425.
71 'Senex' [R. Reid], *Glasgow past and present*, vol. III, p. 83.
72 'Glasgow Highland Society: rules and regulations and list of members 1727–1902' (Glasgow, 1902), pp. 7–10; *Extracts from the records of the Burgh of Glasgow 1809–1822* (Glasgow, 1915), vol. X, p. 698, 18th October 1822; Glasgow University Library, MSS Eph. J/13, Glasgow Society for Support of Gaelic Schools: Statement of 1841, f. 2; *Report of the Committees of Management of the Auxiliary Society of Glasgow for the Support of Gaelic Schools* (Glasgow, 1815), p. 14.
73 D. Smith, *Conflict and compromise*, pp. 6–7.
74 Morris, 'Voluntary societies and British urban élites, 1780–1850: an analysis'.
75 'Report from Her Majesty's Commissioners' (see note 49), p. 642, QQ. 11, 603.
76 'Report from the Select Committee' (see note 50), p. 117, Q. 1241.
77 'Report from Her Majesty's Commissioners', p. 650, Q. 11, 674.
78 Ibid., p. 648, Q. 11, 655.
79 Werly, 'The Irish in Manchester', pp. 348–9.
80 'Report from the Select Committee' (see note 50), p. 47.
81 Lobban, 'The migration of Highlanders into Lowland Scotland'.
82 De Rousier, *The labour question in Britain*, pp. 301–2.

4 The country and the city: sexuality and social class in Victorian Scotland

1 Williams, *The country and the city*, p. 58.
2 Maclaren, *Religion and social class*; Gray, *The labour aristocracy*.
3 Smout, 'Aspects of sexual behaviour'; Boyd, *Scottish church attitudes*.
4 Harvey, *Consciousness and the urban experience*.
5 Doherty, 'Urbanization, capital accumulation and class struggle', p. 243.

6 Weeks, *Sex, politics and society*, p. 57.

7 Williams, *The country and the city*, p. 261.

8 Macgregor, 'Social research and social policy', pp. 147–8.

9 Carter, 'Illegitimate births and illegitimate inferences', p. 125; Scotland's 'national shame' was exaggerated, since English births had been under-registered. See Smout, 'Aspects of sexual behaviour', pp. 62–3; J. M. Strachan, 'Immorality in Scotland', *The Scotsman*, 20 May 1870, p. 6.

10 Free Church General Assembly Proceedings and Debates, 1860, Housing Report, p. 9, cited in Boyd, *Scottish church attitudes*, p. 35.

11 Abrams, *Origins of British sociology*; Murray, *Losing ground*; Ryan, *Blaming the victim*.

12 Chalmers, 'Christian and civic economy', pp. 107–8, cited in Smout, 'Aspects of sexual behaviour', p. 62.

13 Boyd, *Scottish church attitudes*, p. 171.

14 Title of ch. 3 in Henriques, *Modern sexuality*.

15 Teitelbaum, *British fertility decline*, p. 149; Table 6.10b (p. 152).

16 Smout, 'Aspects of sexual behaviour'.

17 Boyd, *Scottish church attitudes*, p. 125.

18 Muirhead, 'Churchmen and the problem of prostitution', p. 223.

19 Free Church General Assembly Proceedings and Debates (1867), Housing Report, p. 7; ibid. (1860) Housing Report, p. 4, cited in Boyd, *Scottish church attitudes*, pp. 39, 33.

20 Macgregor, 'Social research and social policy', p. 148.

21 Begg, *Pauperism and the Poor Laws* notes the threat of communism (p. 44) whilst remarking that; 'A great flight of aristocratic pensioners from above, and a great swarm of paupers and criminals from below, have gradually placed the middle classes between the two fires which equally threaten to consume them.' (pp. 7–8); Seton, *The causes of illegitimacy*, Appendix IV; I. Carter, 'Illegitimate births and illegitimate inferences', p. 126; Cramond, *Illegitimacy in Banffshire*, p. 18. In 1858, according to Cramond, 'there was not a single illegitimate case of a fishergirl or fisherman's daughter' [in Banffshire]. In 1868 there was one; by 1886, twenty-one (Cramond, 'Illegitimacy in Banffshire', pp. 574, 78).

22 'C' [William Cramond], 'Illegitimacy in Banffshire', *The Scotsman*, 13 March 1887.

23 List, *The two phases*, p. 3.

24 Carter, *Farm life*; id., 'Class and culture'; Gray, 'Scottish emigration'; id., 'North-East Agriculture'; id., 'Farm workers'.

25 Begg, 'Houses of the working-classes', p. 624.

26 Cited in Cramond, *Illegitimacy in Banffshire*, p. 41.

27 Kussmaul, *Servants in husbandry*.

28 Levine and Wrightson, 'The social context'.

29 Extensively documented in Boyd, *Scottish church attitudes*, pp. 28–45.

30 Alexander, 'The peasantry of northeast Scotland'.

31 Gerrard, *The rural labourers*, p. 18.

32 Stuart, *Agricultural labourers*, pp. 19–20; Anon., 'The physical condition', pp. 630–1). Cf. Harvey, *Consciousness and the urban experience*, p. 138: 'A contained woman, contained in a corset, contained in a house was an orderly woman.'

33 Cowie et al., 'Digest of essays'; Nicol, *A voice*, p. 8.
34 Anon., 'The condition', p. 731; Begg, 'Obstacles to cottage-building', p. 691.
35 Smout, 'Aspects of sexual behaviour', p. 66.
36 British Parliamentary Papers (BPP) 1870 [C. 70] xiii, p. 1, Royal Commission on the employment of children, young persons and women in agriculture. Third report, p. 8.
37 Boyd, *Scottish church attitudes*, p. 36.
38 Strachan, 'Immorality in Scotland'. Whereas in Ross-shire only 5 per cent of registered working-class births were illegitimate and in only 15 per cent of cases were children born within six months after wedlock, in the Strathbogie district of the northeast the figures were 24 per cent and 65 per cent respectively; See id., *Address upon illegitimacy*, pp. 6–8.
39 Alexander, 'The peasantry of Northeast Scotland' (December 1884), p. 520.
40 Thomas, 'The double standard'.
41 Tait, *Magdalenism*, p. 141; Vacher, *Seduction in Edinburgh* [I am grateful to Kenneth Boyd for supplying me with a copy of this pamphlet].
42 Strachan, 'Immorality in Scotland'.
43 Gray, *The labour aristocracy*; id., 'Bourgeois hegemony'.
44 Goode, 'Illegitimacy', p. 912.
45 Cramond, *Illegitimacy in Banffshire*, p. 44 *et passim*; 'C', 'Illegitimacy in Banffshire'; Cramond, 'Illegitimacy in Banffshire', pp. 575, 583; Charles, *Draft of a report*.
46 Cramond, *Illegitimacy in Banffshire*, pp. 47, 49.
47 Blaikie, thesis, ch. 4.
48 Carter, *Farm life*, *passim*.
49 Ibid., p. 179.
50 Sontag, *AIDS*.
51 Smout, 'Aspects of sexual behaviour', p. 57.
52 Farquhar Spottiswoode, *A plea for emigration*; Blaikie, thesis, ch. 2.
53 List, *The two phases*, pp. 15–17.
54 *Report by a Committee of the Prison Board*, pp. 14–15. This essay predates the discussion of the bourgeois discourse on prostitution provided in Mahood, *The Magdalenes*. See also Finnegan, *Poverty and prostitution*.
55 Maclaren, 'Bourgeois hegemony and Victorian philanthropy', p. 48.
56 Cowan, *Vital statistics of Glasgow*, p. 34.
57 Cited in Mechie, *The church and Scottish social development*, p. 27.
58 List, *The two phases*, p. 14.
59 Seton, *The causes of illegitimacy*, pp. 23–4.
60 Wilson, 'Law of infanticide'.
61 Seton, *The causes of illegitimacy*, Appendix V.
62 *Report by a Committee of the Prison Board*, p. 16; List, *The two phases*, p. 4.
63 Thomson, *Social evils*, pp. 21ff.
64 *Draft Report by Sub-Committee of Rural Police*.
65 *Report by a Committee of the Prison Board*, pp. 7–8.
66 Watson, *Pauperism, vagrancy, crime*, pp. 11–12; id., 'The Poor Law', pp. 1–8.
67 Watson, 'Rural Police in Scotland', p. 315; id., *Pauperism, vagrancy, crime*, p. 1; Thomson, 'A brief history'.

68 Valentine, 'Illegitimacy in Aberdeen'; *Report by a Committee of the Prison Board*, pp. 16–18. (Ironically, in later twentieth-century Aberdeen, 'getting off at Union Terrace' became a working-class colloquialism for coitus interruptus. Union Terrace is the last bus stop before the station.)
69 *Report by a Committee of the Prison Board*, pp. 9–10.
70 Blaikie, thesis, ch. 6.
71 Watson, 'Rural Police in Scotland', p. 315.
72 Watson, *Pauperism, vagrancy, crime*, pp. 1–17.
73 Pringle, 'The agricultural labourer of Scotland', p. 248.
74 Cramond, *Illegitimacy in Banffshire*, p. 43; Alexander, 'Aberdeenshire character', pp. 10–11.
75 Drummond, *Onward and upward*; *Onward and Upward (O&U)* 4:9 (August 1894), p. 213.
76 Countess of Aberdeen, 'What is the Haddo House Association?', *O&U* 1:1 (December 1890), p. 18).
77 *O&U* 5:6 (April 1895), p. 134.
78 Drummond, *Onward and upward*, pp. 4–5; *O&U* 2:3 (February 1892), p. 52; List, *The two phases*, p. 14. Eschatologically based threats also appeared in Anon., *A well meant word*, pp. 2–3 and Synod of Aberdeen, *Pastoral address*, p. 7.
79 *O&U* 4:9 (August 1894), p. 212; 2:2 (January 1892), pp. 45–6.
80 *O&U* 1:12 (November 1891), p. 232; 2:6 (May 1892), pp. 129–30; 2:1 (December 1891), p. 293; 4:10 (September 1894), p. 244.
81 Countess of Aberdeen, 'What is the Haddo House Association?', p. 19.
82 Carter, 'Dorset, Kincardine and peasant crisis', p. 48.
83 Ryan, *Blaming the victim*, pp. 7–8.
84 Cohen, *Folk devils and moral panics*.
85 Anon., *The great sin of Banffshire*, p. 1; Curtis, 'Impurity', p. 72.
86 Blaikie, thesis.
87 Checkland, *Philanthropy in Victorian Scotland*, p. 242.; Weeks, *Sex, politics and society*, p. 84, citing Harrison, 'State intervention and moral reform'.
88 Harvey, *Consciousness and the urban experience*, p. 135; Smout, 'Aspects of sexual behaviour', p. 69; List. *The two phases*, p. 42.
89 List, *The two phases*, p. 42.
90 Blaikie, thesis, chs. 6 and 7.
91 Boyd, *Scottish church attitudes*, p. 181.
92 Weeks, *Sex, politics and society*, p. 32.
93 Boyd, *Scottish church attitudes*, pp. 58–9, 100.
94 Smout, 'Aspects of sexual behaviour', p. 81.
95 Shorter, 'Illegitimacy, sexual revolution and social change'; Scott and Tilly, 'Women's work', pp. 41–2, 62.
96 Weeks, *Sex, politics and society*, p. 62.
97 Begg, *The ecclesiastical and social evils*.
98 Boyd, *Scottish church attitudes*, p. 32; Drummond and Bulloch, *The church in Victorian Scotland*, p. 138.
99 Foucault, *The history of sexuality*.
100 Noble, 'Urbane silence', pp. 64, 73, 79–80.

5 Mobility, the artisan community, and popular politics in early nineteenth-century England

1 Hobsbawm, 'The tramping artisan', p. 34. Page numbers refer to 1964 reprint.
2 Well-known studies of towns include: Foster, *Class struggle* (Oldham); Joyce, *Work, society and politics* (Blackburn); studies emphasising artisans are: Crossick, *An artisan élite* (Woolwich engineers); Gray, *The labour aristocracy* (Edinburgh printers); Matsumura, *Flint glass makers* (Stourbridge). For workplace-centred studies, see: Price, *Master, unions, and men*; Harrison and Zeitlin, *Divisions of labour*; Joyce, *Historical meaning of work*.
3 Friedlander and Roshier, 'A study of internal migration'; Baines, *Migration in a mature economy*.
4 Hunt, *Regional wage variations*, ch. 7.
5 Lees, *Exiles of Erin*; Pooley, 'The residential segregation'; Anderson, *Family structure*.
6 Garraty, *Unemployment in history*, pp. 12–13.
7 Leeson, *Travelling brothers*, pp. 92–3; Rule, *The experience of labour*, pp. 164–5.
8 Leeson, *Travelling brothers*, pp. 59–78, 90; Dobson, *Masters and journeymen*, pp. 47–59.
9 Southall, 'Geography of unionization', pp. 467–72.
10 Webb and Webb, *Trade unionism*, pp. 23–4; Leeson, *Travelling brothers*, pp. 93–5; Randall, 'Wiltshire outrages', p. 290.
11 Leeson, *Travelling brothers*, p. 112.
12 In addition to earlier references, see Hobsbawm, 'The tramping artisan', pp. 39–42; Musson, *The Typographical Association*, pp. 24–7, 50–2.
13 For a description of the system, see Leeson, *Travelling brothers*, pp. 122–47. For the Steam Engine Makers' Society, see Southall, 'The tramping artisan revisits'.
14 Southall, 'Regional unemployment patterns', thesis, pp. 67, 71.
15 Leeson, *Travelling brothers*, pp. 211–39.
16 Burnett et al., *Autobiography of the working class*, col. 1, covering authors born before 1900; this indexes sixty-one entries under 'tramping'. Of the nine unread autobiographies, the only copies of five are in impractical locations and four were by men born after 1875 who say, from the abstracts, little or nothing about tramping in Britain. The only woman among the sixty-one was Julia Varley (973), and she was a Poor Law guardian who tramped in disguise in the 1900s as part of a social investigation. Given that many are inaccessible, and to save space, the autobiographies are referenced here by the name of the author and the number of the main entry in the bibliography.
17 Thompson, *The Chartists*, p. 92.
18 Hobsbawm, 'History from below', p. 17.
19 Miles Watkins (88), p. 1.
20 Burnett et al., *Autobiography of the working class*, pp. xxvi–xxvii.
21 James Henry Powell (561), p. 10.
22 James Dunn (216), p. 22.
23 Southall, 'The tramping artisan revisits'.
24 Robert Gammage (261). This set of 'reminiscences', rather than an auto-biography, originally appeared in 1883–5 as a series of seventeen articles in the

Newcastle Weekly Chronicle, edited by William Adams, the author of another autobiography used here. Although they cover certain incidents, particularly some of his travels, in considerable detail, they were not organised chronologically and make no attempt to document his life as a whole, additional details being provided by the editor of the modern reprint: Maehl, *Robert Gammage.*

25 Ibid., p. 48.

26 The Chartist missionary Henry Vincent, who influenced Gammage in Northampton, was attacked by a hostile mob in Devizes the previous year: Bellamy and Saville, *Dictionary of labour biography*, vol. 6, p. 115; id., vol. 1, p. 327.

27 Robert Gammage (261), p. 49.

28 Ibid., p. 51.

29 Ibid., pp. 26–9.

30 William Adams (5), pp. 171–2.

31 For some inconclusive calculations of occupation-specific lifetime mobility from the census, see Southall, 'The tramping artisan revisits'.

32 The two tanners' rule books are in a bound collection in the British Library under the title 'Men's permanent societies' (8275.bb.4); a map of their branches is given in Southall, 'Towards a geography of unionization', p. 476. All figures for the SEM from Southall, 'The tramping artisan revisits'.

33 Henry Broadhurst (88), p. 11.

34 Quotes from: William Fairbairn (234), p. 25; Henry Herbert (324), p. 54; James Powell (561), p. 10; Robert Spurr (655), p. 283. See also J.E. (218), p. 3; John Bedford Leno (430), p. 24; Charles Manby Smith (632), p. 13; Joseph Stamper (658), p. 7; Thomas Wood (772), 7/4/1856, p. 7; Thomas Wright (897), p. 255.

35 J.E. (218), p. 3; Paul Evett (233), p. 3; William Farish (236), p. 223; Alfred Ireson (371), p. 47; Jacques (375), 1/11/1856, p. 3; Anon. (436), p. 352; George Mitchell (515), p. 109; Thomas Preston (563), pp. 5, 7–8; John Savage (607), p. 2; Stir(r)up (667), 6/12/1856, p. 2; quote from J.G. (253), p. 7.

36 Thomas Wood (772), 7/4/1856, p. 7.

37 George Rowles (600), p. 37; Thomas Wallis (726), p. 25; Miles Watkins (734), p. 22; Thomas Wright (897), p. 271.

38 Thomas Wright (897), p. 161.

39 Robert Gammage (261), p. 8; James Hillocks (328), p. 37; Anon. (436), p. 315; Thomas Preston (563), p. 15; Edward Rymer (603), p. 12; Anon. (995), 13/7/1850, pp. 20–1 (copy in Goldsmiths' Library, London).

40 George Herbert (323), p. 12.

41 Miles Watkins (734), pp. 25–30, 36–8.

42 George Meek (505), p. 46; Will Thorne (708), p. 30.

43 John Buckley (101), p. 113; Bill H. (292), pp. 11–12; William Hutton (366), p. 30; Frank Kitz (417), 1/1912, p. 2.

44 Henry Broadhurst (88), p. 20; see also Josiah Bassett (50), p. 17; J.E. (218), p. 3; Frank Kitz (417), 1/1912, p. 2; Thomas Wright (897), pp. 259–60, 262.

45 William Adams (5), p. 304; William Fairbairn (234), p. 99; William Farish (235), p. 25; William Hutton (366), p. 72.

46 Robert Gammage (261), p. 72.

47 William Adams (5), p. 299; Josiah Bassett (50), p. 24; George Holyoake (344), p. 71.

48 William Marcroft (486), pp. 6–8.
49 J.E. (218), p. 3; Thomas Wallis (726), p. 14.
50 J.E. (218), p. 3; William Fairbairn (234), p. 85; William Farish (236), p. 224; J.G. (253), p. 7; Bill H. (292); Thomas Preston (563), p. 5; John Savage (607), p. 5; Robert Spurr (655), p. 283; Thomas Wood (772), 7/4/1856, p. 7 (although he disliked the fellow apprentice, who often walked half a mile ahead!); Thomas Wright (897), p. 261.
51 Josiah Bassett (50), p. 12; William Farish (235), p. 25; Bill H. (292), p. 13; Alfred Ireson (371), p. 54; John Bedford Leno (430).
52 William Adams (5), p. 301.
53 Henry Broadhurst (88), p. 24.
54 William Adams (5), p. 46.
55 Fred Bower (78), ch. 4; see also 'Colin' (164), p. 51; William Fairbairn (234), p. 101; George Herbert (323), p. 12; Alfred Ireson (371), p. 70; Thomas Preston (563), p. 11.
56 George Herbert (323), p. 18.
57 Ibid., p. 18.
58 William Fairbairn (234), p. 88; T.J. Hunt (362), p. 16; Alfred Ireson (371), p. 51; John Bedford Leno (430), p. 50; Edward Rymer (603), p. 7; Charles Manby Smith (632), p. 141.
59 Henry Broadhurst (88), p. 24; William Fairbairn (234), p. 92; William Farish (236), p. 227; Anon. (436), p. 375; Robert Scott (613), p. 25; Will Thorne (708), pp. 49, 50, 53; Thomas Wood (772), 14/4/1856, p. 7; Thomas Wright (897), p. 281.
60 Alfred Ireson (371), p. 58, lodging with other Wesleyans met in Gloucester, is the only example.
61 Thomas Wright (897), pp. 257–8.
62 Henry Broadhurst (88), p. 18.
63 William Adams (5), p. 292; Bill H. (292), pp. 14, 52; George Herbert (323), p. 13; Albert Pugh (565), p. 89; Will Thorne (708), p. 34.
64 Thomas Wright (897), p. 259.
65 Henry Broadhurst (88), p. 21.
66 Frank Kitz (417), 1/1912, p. 2; Albert Pugh (565), pp. 86–7; Joseph Stamper (658), *passim.*; Will Thorne (708), p. 30.
67 William Adams (5), pp. 292, 298.
68 Henry Broadhurst (88), p. 22. For other examples of men using clubhouses, see 'Colin' (164), p. 31; J.E. (218), p. 3; Alfred Ireson (371), pp. 57, 72; James Powell (561), pp. 12, 33, 34; Stir(r)up (667), 6/12/1856, p. 3.
69 John Buckley (101), p. 115. Buckley is a pseudonym.
70 Ibid., p. 114.
71 Ibid., pp. 116–18.
72 George Herbert (323), p. 21.
73 Paul Evett (233), p. 5.
74 Will Thorne (708), p. 36.
75 George Rowles (600), p. 69.
76 Bill H. (292), pp. 41–4.
77 Kussmaul, *Servants in husbandry*, ch. 4; Armstrong, *Farmworkers*, p. 22.

78 Edward Rymer (603), p. 8, writing in 1898.
79 'Colin' (164), p. 28; J. E. (218), p. 3; Alfred Ireson (371), p. 56; Charles Manby Smith (632), p. 17.
80 Alfred Ireson (371), p. 63; James Powell (561), pp. 34, 40.
81 William Farish (236), p. 224.
82 George Rowles (600), pp. 63. See also Henry Herbert (324), p. 62.
83 William Fairbairn (234), p. 91; George Meek (505), p. 54; Thomas Preston (563), p. 9.
84 Thomas Wright (897), p. 281, see also p. 160; Henry Broadhurst (88), p. 23; Alfred Ireson (371), pp. 72–3.
85 George Rowles (600), p. 64.
86 William Marcroft (486), p. 11.
87 John Buckley (101), p. 115; Peter Gabbitass (253A), p. xviii; Alfred Ireson (371), pp. 50–1; Thomas Wallis (726), p. 54; Miles Watkins (734), pp. 27, 38.
88 Alfred Ireson (371), p. 77. See also Jacques (375), 1/11/1856, p. 3.
89 An exception is Bill H. (292), pp. 40–6.
90 Edward Rymer (603), p. 9. The peculiar nature of mining unionism is a central theme of Southall, 'Towards a geography of unionization'.
91 See also Fred Bower (78), Henry Broadhurst (88), Will Thorne (708).
92 Bellamy and Saville, *Dictionary of labour biography*, vol. 1, p. 14.
93 Ibid., vol. 2 (1974), p. 270.
94 Ibid., p. 16.
95 Epstein, *The lion of freedom*, pp. 90–1; Thompson, *The Chartists*, p. 105.
96 E. P. Thompson, *The making of the English working class*, pp. 111–203.
97 Ibid., pp. 685–7. See also Wells, *Insurrection*, pp. 171–2, on Cartwright's earlier activities.
98 Thompson, *The making of the English working class*, pp. 679–90, quote from p. 689; Prothero, *Artisans and politics*, pp. 120–1.
99 Thompson, *The Chartists*, pp. 96–101, quote from p. 99.
100 The main source is the *Northern Star*; the figure attempts to show all public speeches but excludes committee meetings, numerous court appearances and attendance at the Chartist National Convention. O'Connor's house was in Hammersmith, London, and the *Northern Star* offices were in Leeds; hence he spent much more time in these two towns, and travelling between them, than the map suggests.
101 *Northern Star*, 29 September 1838, p. 4.
102 *Northern Star*, 19 January 1839, p. 4.
103 *Northern Star*, 21 July 1838, p. 3; see also ibid., 4 August 1838, p. 6, 29 December 1838, p. 8, 22 June 1839, p. 8.
104 Epstein, *The lion of freedom*, pp. 32–3, 35, 38, 48–9, 52, 110, 111, 131.
105 Harrison and Hollis, *Robert Lowery*, pp. 107–17.
106 Thompson, *The Chartists*, pp. 133, 175, and *passim*; Bellamy and Saville, *Dictionary of labour biography*, vol. 1, pp. 326–34.
107 Jenkins, *The general strike*, pp. 120–1; Bellamy and Saville, *Dictionary of labour biography*, vol. 6, pp. 216–23.
108 Josiah Bassett (50), p. 55.
109 William Farish (235), pp. 36–9 on radicalism, p. 62 on temperance; see also

Stir(r)up (667), 29/11/1856, p. 3; Anon. (995), 13/7/1850, p. 21 (copy in Goldsmiths' Library, London).

110 Thompson, *The making of the English working class*, pp. 678, 705, 713–17 (quote from p. 713).

111 Ibid., pp. 713, 719, 910; Bellamy and Saville, *Dictionary of labour biography*, vol. 6, pp. 29–36; Prothero, *Artisans and politics, passim*; Thompson, *The Chartists*, p. 279.

112 William Adams (5), pp. 168–72, 288–9.

113 John Buckley (101), pp. 124, 137, 156.

114 James Hillocks (328), p. 37.

115 John Bedford Leno (430), pp. 19, 28.

116 Anon. (436), p. 399.

117 Thomas Preston (563), pp. 5, 8, 10.

118 Prothero, *Artisans and politics*, p. 16.

119 E.g. Donelly and Baxter, 'Sheffield and the English revolutionary tradition'; Epstein and Thompson, *The Chartist experience*.

120 P. Anderson, *Arguments within English Marxism*, p. 31; see also McLennan, 'E. P. Thompson', p. 110.

121 Langton, 'The industrial revolution', esp. pp. 150–5; for a critique of Langton's comments on the union movement, see Southall, 'Towards a geography of unionization'.

122 Thompson, *The makings of the English working class*, pp. 913–14.

Consolidated bibliography

(Unless otherwise stated, place of publication is London.)

Aberdeen, Countess of, 'What is the Haddo House Association?', *Onward and Upward* 1:1 (December 1890), pp. 18–22.

Aberdeen, Synod of, *Pastoral address on the moral condition of the rural population* (Aberdeen, 1865).

Abrams, P., *The origins of British sociology, 1834–1914* (Chicago, Ill., 1968).

Adamson, W., *Hegemony and revolution* (Berkeley, Calif., 1980).

Agnes, J., Mercer, J., and Sopher, D., eds., *The city in cultural context* (Boston, Mass., 1984).

Alexander, W., 'The peasantry of northeast Scotland', *United Presbyterian Magazine* NS 1 (September 1884), pp. 377–9; (October 1884), pp. 426–9; (December 1884), pp. 519–23.

 'Aberdeenshire character and characteristics', *Onward and Upward* 2:1 (1891), pp. 10–15.

Anderson, B., *Imagined communities: reflections on the origin and spread of nationalism* (1983).

Anderson, M., *Family structure in nineteenth-century Lancashire* (Cambridge, 1971).

Anderson, P., 'The antinomies of Antonio Gramsci', *New Left Review* 100 (1976–7), pp. 5–80.

 Arguments within English Marxism (1980).

Armstrong, A., *Farmworkers: a social and economic history 1770–1980* (1988).

Armstrong, W. A., 'The use of information about occupation' in Wrigley, E. A., ed., *Nineteenth-century society: essays in the use of quantitative methods for the study of social data* (Cambridge, 1972), pp. 191–310.

Aspinwall, B., and McCaffrey, J., 'A comparative view of the Irish in Edinburgh in the nineteenth century' in Swift, R., and Gilley, S., eds., *The Irish in the Victorian city* (1985), pp. 130–57.

Badcock, B., *Unfairly structured cities* (Oxford, 1984).

Bailey, P., *Leisure and class in Victorian England: rational recreation and the contest for control, 1830–85* (1978).

 '"Will the real Bill Banks please stand up?" Towards a role analysis of mid-Victorian respectability', *Journal of Social History* 12 (1979), pp. 336–53.

Baines, D., *Migration in a mature economy: emigration and internal migration in England and Wales 1861–1900* (Cambridge, 1985).

Bairoch, P., *Cities and economic development from the dawn of history to the present* (1988).

Bates, T. R., 'Gramsci and the theory of hegemony', *Journal of the History of Ideas* 36 (1975), pp. 351–66.

Begg, J., *Pauperism and the Poor Laws, or our sinking population and rapidly increasing public burdens practically considered* (2nd edn, Edinburgh, 1849).

'Houses of the working-classes of Scotland: the bothy system', *Transactions of the National Association for the Promotion of Social Science* (1858), pp. 621–4.

'Obstacles to cottage-building in Scotland, and how to remove them', *Transactions of the National Association for the Promotion of Social Science* (1859), pp. 690–6.

The ecclesiastical and social evils of Scotland, and how to remedy them (Edinburgh, 1871).

Behagg, C., 'Secrecy, ritual and folk violence: the opacity of the workplace in the first half of the nineteenth century' in Storch, R. D., ed., *Popular culture and custom in nineteenth-century England* (1982), pp. 154–79.

Bellamy, J., and Saville, J., eds., *Dictionary of labour biography*, vols. I–VII (1972–82).

Beresford, M. W., 'Prosperity street and others: an essay in visible urban history' in Beresford, M. W. and Jones, G. R. J., eds., *Leeds and its region* (Leeds, 1967), pp. 168–97.

East End, West End: the face of Leeds during industrialisation 1684–1842 (Leeds, 1988).

la Berge, A. F., 'Edwin Chadwick and the French connection', *Bulletin of the History of Medicine* 62 (1988), pp. 23–42.

Berry, B. J. L., 'Cities as systems within systems of cities', *Papers and Proceedings of the Regional Science Association* 13 (1967), pp. 147–63.

Billinge, M. D., 'Hegemony, class and power in later Georgian and early Victorian England: towards a cultural geography' in Baker, A. R. H., and Gregory, D. J., eds., *Explorations in historical geography* (Cambridge, 1984), pp. 28–67.

Billinge, M. D., Gregory, D., and Martin, R. L., eds., *Recollections of a revolution: geography as spatial science* (1984).

Blaikie, J. A. D., 'Illegitimacy in nineteenth century northeast Scotland' (unpublished Ph.D. thesis, University of London, 1987).

Blaut, J. M., 'A radical critique of cultural geography', *Antipode* 12 (1980), pp. 25–9.

Boyd, K., *Scottish church attitudes to sex, marriage and the family, 1850–1914* (Edinburgh, 1980).

Briggs, A., *A history of Birmingham, vol. II: borough and city 1865–1938* (Oxford, 1952).

'Cholera and society in the nineteenth century', *Past and Present* 19 (1961), pp. 76–96.

Victorian cities (1963).

Brindley, J. M., *Church work in Birmingham*, vol. I (Birmingham, 1878); vol. II (Birmingham, 1885).

Brookfield, H. C., *Interdependent development* (1973).

Brown, C. G., 'The costs of pew-renting: church management, church-going and social class in nineteenth-century Glasgow', *Journal of Ecclesiastical History* 38:3 (1987), pp. 347–61.

Buci-Glucksmann, C., *Gramsci and the state* (1980).

Buckman, J., 'Alien working-class response: the Leeds Jewish tailors, 1880–1914' in Lunn, K., ed., *Hosts, immigrants and minorites: historical response to newcomers in British society 1870–1914* (Folkestone, 1980), pp. 222–62.

Bunce, J. T., *History of the corporation of Birmingham*, vol. II (Birmingham, 1885).

Burnett, J., Vincent, D., and Mayall, D., *Autobiography of the working class: an annotated critical bibliography*, vol. I (1790–1900) (Brighton, 1984).

Bushaway, R., *By rite: custom, ceremony and community in England, 1700–1880* (1982).

'C' [William Cramond], 'Illegitimacy in Banffshire', *The Scotsman*, 13 March 1887, p. 5.

Cage, R. A., 'The standard of living debate: Glasgow, 1800–1850', *Journal of Economic History* 43:1 (1983), pp. 175–82.

    ed., *Working-class in Glasgow, 1750–1914* (1984).

Calhoun, C., 'Community: toward a variable conceptualisation for comparative research', *Social History* 5 (1980), pp. 105–29.

    'Class, place and the industrial revolution' in Thrift, N., and Williams, P., eds., *Class and space* (1987), pp. 51–72.

Canetti, E., *Crowds and power* (1962).

Cannadine, D., 'Victorian cities: how different?', *Social History* 2 (1977), pp. 457–82.

    'Residential differentiation in nineteeth-century towns: from shapes on the ground to shapes in society' in Johnson, J. and Pooley, C. G., eds., *The structure of nineteenth-century cities* (1982), pp. 235–51.

Carter, H., *An introduction to urban historical geography* (1983).

Carter, H., and Wheatley, S., 'Some aspects of the spatial structure of two Glamorgan towns in the nineteenth century', *Welsh History Review* 9 (1978), pp. 32–56.

Carter, I., 'Dorset, Kincardine and peasant crisis: a comment on David Craig', *Journal of Peasant Studies* 2:4 (1975), pp. 151–91.

    'Class and culture among farm servants in the north-east, 1840–1914' in MacLaren, A. A., ed., *Social class in Scotland: past and present* (Edinburgh, 1976), pp. 105–27.

    'Illegitimate births and illegitimate inferences', *Scottish Journal of Sociology* 1:2 (1977), pp. 125–35.

    Farm life in northeast Scotland: the poor man's country (Edinburgh, 1979).

Castells, M., *The urban question: a Marxist approach* (1977).

    City, class and power (1978).

Chalklin, C. W., *The provincial towns of Georgian England: a study of the building process, 1740–1820* (1974).

Chalmers, T., *The Christian and civic economy of large towns*, vol. III, in *Works* (Glasgow, 1838–42), vol. XVI (cited in Smout, 1976).

Charles, J., *Draft of a report anent illegitimacy given to the Presbytery of Wigtown* (Glasgow, 1864).

Cheal, D., 'Hegemony, ideology and contradictory consciousness', *The Sociological Quarterly* 20 (1979), pp. 109–17.

Checkland, O., *Philanthropy in Victorian Scotland: social welfare and the voluntary principle* (Edinburgh, 1980).

Chevalier, L., *Labouring classes and dangerous classes in Paris during the first half of the nineteenth century* (1973; original French edition, 1958).

Clark, P., and Souden, D., eds., *Migration and society in early modern England* (1987).

Cohen, S., *Folk devils and moral panics* (1972).

Coleman, B. I., *The idea of the city in the nineteenth century* (1973).

Coleman, W., *Death is a social disease: public health and political economy in early industrial France* (1982).

Collins, B., 'The social experiences of Irish migrants in Dundee and Paisley during the mid-nineteenth century', paper presented to the Urban History Conference (Sheffield, 1979).

'The condition of the agricultural labourer', *Journal of Agriculture* 22 (1859–61), pp. 721–36.

Connell, J., 'The gilded ghetto: Jewish suburbanisation in Leeds', *Bloomsbury Geographer* 3 (1970), pp. 50–9.

Conzen, K., 'Immigrants, immigrant neighbourhoods and ethnic identity: historical issues', *Journal of American History* 66 (1979), pp. 603–15.

Corrigan, P., and Sayer, D., *The great arch: English state formation as cultural revolution* (Cambridge, 1985).

Cosgrove, D., 'Towards a radical cultural geography', *Antipode* 15 (1983), pp. 1–11.

Cowan, R., *Vital statistics of Glasgow* (Glasgow, 1840).

Cowie, J., et al., 'Digest of essays on the bothy system of maintaining single servants', *Transactions of the Highland and Agricultural Society of Scotland* NS 14 (1843), pp. 133–44.

Cowlard, K. A., 'The urban development of Wakefield, 1801–1901' (unpublished Ph.D. thesis, University of Leeds, 1974), pp. 35–41.

Cramond, W., *Illegitimacy in Banffshire: facts, figures and opinions* (Banff, 1888).
'Illegitimacy in Banffshire', *Poor Law Magazine* NS (1892).

Crossick, G., *An artisan élite in Victorian society: Kentish London 1840–1880* (1978).

Cunningham, H., 'The metropolitan fairs: a case study in the social control of leisure', in Donajgrodski, A. P., ed., *Social control in nineteenth-century Britain* (1977), pp. 163–84.

Curtis, W. A., 'Impurity' in Paterson, W. P., and Watson, D., eds., *Social evils and problems* (Edinburgh, 1918), pp. 65–91.

Darby, H. C., 'The movement of population to and from Cambridgeshire between 1851 and 1861', *Geographical Journal* 101 (1943), pp. 118–25.

Daunton, M. J., 'The building cycle and the urban fringe in Victorian cities: a comment', *Journal of Historical Geography* 4 (1978), pp. 175–81.
House and home in the Victorian city: working-class housing, 1850–1914 (1983).
'Public place and private space: the Victorian city and the working-class household' in Fraser, D., and Sutcliffe, A., eds., *The pursuit of urban history* (1983), pp. 212–33.

Davidoff, L., and Hall, C., 'The architecture of public and private life: English middle-class society in a provincial town, 1760 to 1850' in Fraser, D., and Sutcliffe, A., eds., *The pursuit of urban history* (1983), pp. 327–45.

Delaporte, F., *Disease and civilisation: the cholera in Paris, 1832* (Cambridge, Mass., 1986).

Dennis, R., and Prince, H., 'Research in British urban historical geography' in Denecke, D., and Shaw, G., eds., *Urban historical geography: recent progress in Britain and Germany* (Cambridge, 1988), pp. 9–23.

Dennis, R. J., 'Community and interaction in a Victorian city: Huddersfield, 1850–1880' (unpublished Ph.D. thesis, University of Cambridge, 1975).

English industrial cities of the nineteenth century: a social geography (Cambridge, 1984).

Dent, R. K., *The making of Birmingham* (Birmingham, 1894).

Dentith, S., 'Political economy, fiction and the language of practical ideology in nineteenth-century England', *Social History* 8:2 (1983), pp. 183–99.

Devine, T. M., 'Temporary migration and the Scottish Highlands in the nineteenth century', *Economic History Review* 33 (1979), pp. 344–59.

'Highland migration to Lowland Scotland, 1760–1860', *Scottish Historical Review* 62 (1983), pp. 137–49.

The Great Highland famine (Edinburgh, 1988).

Dillon, T., 'The Irish in Leeds, 1851–1861', *Thoresby Society Miscellany* 16 (1974), pp. 1–28.

'The directory of Highland and Celtic societies', *The Celtic Magazine* 4 (1879), pp. 35–8.

Dobson, C. R., *Masters and journeymen: a prehistory of industrial relations 1717–1800* (1980).

Doherty, J., 'Urbanization, capital accumulation and class struggle in Scotland, 1750–1914' in Whittington, G., and Whyte, I. D., eds., *An historical geography of Scotland* (1983), pp. 239–67.

Donajgrodski, A. P., ed., *Social control in nineteenth-century Britain* (1977).

Donelly, F. K., and Baxter, J. L., 'Sheffield and the English revolutionary tradition, 1791–1820', *International Review of Social History* 20 (1975), pp. 398–423.

Draft report by sub-committee of rural police of the County of Aberdeen on the causes of the recent increase of vagrancy and crime in the county (Aberdeen, 1852).

Driver, F., 'Moral geographies: social science and the urban environment in mid-nineteenth century England', *Transactions, Institute of British Geographers* NS 13 (1988), pp. 275–87.

Drummond, A. L., and Bulloch, J. S., *The church in Victorian Scotland, 1843–1874* (Edinburgh, 1975).

Drummond, J., *Onward and upward* (Aberdeen, 1983).

Duncan, J. S., 'The superorganic in American cultural geography', *Annals, Association of American Geographers* 70 (1980), pp. 181–98.

Duncan, O. D., and Duncan, B., *The negro population of Chicago: a study of residential succession* (Chicago, 1957).

Dupré, L., *Marx's social critique of culture* (New Haven, Conn., 1983).

Durey, M., *The return of the plague: British society and the cholera 1831–2* (Dublin, 1979).

Dyos, H. J., *Victorian suburb: a study of the growth of Camberwell* (Leicester, 1966).

Dyos, H. J., and Reeder, D. A., 'Slums and suburbs' in Dyos, H. J., and Wolff, M., eds., *The Victorian city: images and realities*, vol. I (1973), pp. 359–86.

Dyos, H. J., and Wolff, M., eds., *The Victorian city: images and realities* (1973).

Edwards, J., *Language, society and identity* (Oxford, 1985).

Epstein, J., *The lion of freedom: Feargus O'Connor and the Chartist movement, 1832–42* (1982).

Epstein, J., and Thompson, D., eds., *The Chartist experience: studies in working-class radicalism and culture, 1830–60* (1982).

Extracts from the Records of the Burgh of Glasgow 1809–1822 (Glasgow, 1915).

Farquhar Spottiswoode, Mrs, *A plea for emigration in connection with our female industrial schools* (Aberdeen, n.d.).

Feldman, D., 'There was an Englishman, an Irishman and a Jew…: immigrants and minorities in Britain', *Historical Journal* 26:1 (1983), pp. 185–99.

Femia, J., 'Hegemony and consciousness in the thought of Antonio Gramsci', *Political Studies* 23 (1975), pp. 29–48.

Fife, N., and Power, C., *Black settlers in Britain 1555–1958* (1981).

Finer, S. E., *The life and times of Sir Edwin Chadwick* (1952).

Finigan, T. A., *Journal of the Birmingham Town Mission missionary* (1837–8).

Finnegan, F., *Poverty and prostitution: a study of Victorian prostitutes in York* (Cambridge, 1979).

Poverty and prejudice: a study of Irish immigrants in York, 1840–1875 (Cork, 1982).

'The Irish in York' in Swift, R., and Gilley, S., eds., *The Irish in the Victorian city* (1985), pp. 59–84.

Flinn, M. W., ed., *Report on the sanitary condition of the labouring population of Gt Britain by Edwin Chadwick 1842* (Edinburgh, 1965).

Forgacs, D., and Nowell-Smith, G., *Antonio Gramsci – selections from cultural writings* (1983).

Foster, J., *Class struggle and the industrial revolution: early industrial capitalism in three English towns* (1974).

Foucault, M., *The history of sexuality*, vol. I, (1985).

Fraser, D., and Sutcliffe, A., 'Introduction' in Fraser, D., and Sutcliffe, A., eds., *The pursuit of urban history* (1983).

Free Church General Assembly Proceedings and Debates (1860), Housing Report.

Free Church General Assembly Proceedings and Debates (1867), Housing Report.

Fridlizius, G., 'Sex-differential mortality and socio-economic change, Sweden 1750–1910' in Brändström, A., and Tedebrand, L.-G., eds., *Society, health and population during the demographic transition* (Stockholm, 1988), pp. 237–72.

Friedlander, D., and Roshier, R. J., 'A study of internal migration in England and Wales, Part I', *Population Studies* 19 (1966), pp. 239–79.

Gardner, P., *The lost elementary schools of Victorian England: the people's education* (1984).

Garraty, J. A., *Unemployment in history: economic thought & public policy* (New York, 1978), pp. 12–13.

Gaskell, S. M., 'Gardens for the working class: Victorian practical pleasure', *Victorian Studies* 23 (1980), pp. 479–501.

Gay, P., *The bourgeois experience, Victoria to Freud. Volume I: Education of the senses* (Oxford, 1984).

Gerrard, J., *The rural labourers of the north of Scotland: their medical relief and house accommodation as they affect pauperism and illegitimacy* (Banff, n.d. [1862]).

Giglioli, P. P., ed., *Language and social context* (1972).

Gilbert, A. D., *Religion and society in industrial England: churches, chapel and social change, 1740–1914* (1976).

Gilbert, E. W., 'Pioneer maps of health and disease in England', *Geographical Journal* 124 (1958), pp. 172–83.

Gill, C., *History of Birmingham, Volume I: manor and borough to 1865* (Oxford, 1952).

Gilley, S., 'English Catholic charity and the Irish poor in London, Part I, 1700–1814', *Recusant History* 11:4 (1972), pp. 179–95; 'English Catholic charity and the Irish poor in London, Part II, 1840–1870', *Recusant History* 11:5 (1972), pp. 253–69.

'The Catholic faith of the Irish slums' in Dyos, H. J., and Wolff, M., eds., *The Victorian city: images and realities* (1973), pp. 837–53.

'English attitudes to the Irish in England, 1789–1900' in Holmes, C., ed., *Immigrants and minorities in British society* (1978), pp. 81–110.

Gillies, W., ed., *Gaelic and Scotland* (Edinburgh, 1989).

Goode, W. J., 'Illegitimacy, anomie and cultural penetration', *American Sociological Review* 26 (1961), pp. 910–25.

Gourvish, T. R., 'The cost of living in Glasgow in the early nineteenth century', *Economic History Review* 25:1 (1972), pp. 65–80.

Gramsci, A., *Prison notebooks* (1970).

Gray, M., 'Scottish emigration: the impact of agrarian change in the rural Lowlands, 1775–1875', *Perspectives in American History* 7 (1973), pp. 95–174.

'North-east agriculture and the labour force' in MacLaren, A. A., ed., *Social class in Scotland: past and present* (Edinburgh, 1976), pp. 86–104.

'Farm workers in north-east Scotland', in Devine, T. M., ed., *Farm servants and labour in Lowland Scotland 1770–1914* (Edinburgh, 1985), pp. 10–28.

Gray, R. Q., *The labour aristocracy in Victorian Edinburgh* (Oxford, 1976).

'Bourgeois hegemony in Victorian Britain' in Bloomfield, J., ed., *Class, hegemony and party* (1977), pp. 73–93.

The great sin of Banffshire (Banff, 1859).

Gregory, D., *Ideology, science and human geography* (1978).

'Contours of crisis: sketches for a geography of class struggle in the early Industrial Revolution in England' in Baker, A. R. H., and Gregory, D., eds., *Explorations in historical geography* (Cambridge, 1984), pp. 68–117.

Gregory, D., and Urry, J., eds., *Social relations and spatial structures* (1985).

Hamlin, C., 'Providence and putrefaction: Victorian sanitarians and the natural theology of health and disease', *Victorian Studies* 28 (1985), pp. 381–411.

Hammerton, E., and Cannadine, D. N., 'Conflict and consensus on a ceremonial occasion: the diamond jubilee in Cambridge in 1897', *Historical Journal* 24 (1981), pp. 24–40.

Harris, R., 'Residential segregation and class formation in the capitalist city', *Progress in Human Geography* 8 (1984), pp. 26–49.

Harrison, B., *Drink and the Victorians: the temperance question in England 1815–1872* (1971).

'State intervention and moral reform in nineteenth-century England' in Hollis, P., ed., *Pressure from without in early Victorian England* (1974), pp. 289–322.

Harrison, B., and Hollis, P., eds., *Robert Lowery: radical and Chartist* (1979).

Harrison, M., 'The ordering of the urban environment: time, work and the occurrence of crowds, 1790–1835', *Past and Present* 110 (1986), pp. 134–68.

Harrison, R., and Zeitlin, J., eds., *Divisions of labour: skilled workers and technological change in nineteenth-century Britain* (Brighton, 1985).

Harvey, D., *Social justice and the city* (1973).

Consciousness and the urban experience: studies in the history and theory of capitalist urbanization, Volume 2 (Oxford, 1985).

The condition of postmodernity (Oxford, 1989).

Hauser, P. M., and Duncan, O. D., *Methods of urban analysis: a summary report* (San Antonio, Tex., 1955).

Heath, S., *The sexual fix* (1982).

Held, D., and Thompson, J. B., eds., *Social theory of modern societies: Anthony Giddens and his critics* (Cambridge, 1989).

Hennock, E. P., *Fit and proper persons: ideal and reality in nineteenth-century urban government* (1973).

Henriques, F., *Modern sexuality*, vol. III of *Prostitution and society* (1968).

Hepburn, A. C., ed., *Minorities in history* (1978).

Hershberg, T., ed., *Philadelphia: work, space, family and group experience in the nineteenth century: essays towards an interdisciplinary history of the city* (New York, 1981).

Hoare, Q., and Nowell-Smith, G., eds., *Selections from the prison notebooks of Antonio Gramsci* (1971).

Hobsbawm, E. J., 'The tramping artisan', *Economic History Review*, 2nd series, 3 (1951), pp. 299–320; reprinted with a postscript in *Labouring men* (1964).

Labouring men: studies in the history of labour (1964).

'History from below – some reflections', pp. 13–27 in Krantz, ed., *History from below* (Oxford, 1988).

Holmes, C., ed., *Immigrants and minorities in British society* (1978).

'The German gypsy question in Britain, 1904–1906' in Lunn, K., ed., *Hosts, immigrants and minorities: historical responses to newcomers in British society 1870–1914* (Folkestone, 1980), pp. 134–59.

Howe, G. M., *Man, environment and disease in Britain: a medical geography of Britain through the ages* (New York, 1972).

Hughes, H. S., *Consciousness and society: the reorientation of European social thought 1890–1930* (1959).

Humphries, S., *Hooligans or rebels? an oral history of working-class childhood and youth, 1889–1939* (Oxford, 1981).

Hunt, E. H., *Regional wage variations in Britain, 1850–1914* (Oxford, 1973).

'Wages' in Langton, J., and Morris, R. J., eds., *Atlas of industrialising Britain, 1780–1914* (1986), pp. 60–8.

Inglehart, R. F., and Woodward, M., 'Language conflicts and political community' in Giglioli, P. P., ed., *Language and social context* (1972), pp. 358–77.

Inglis, K. S., *Churches and the working classes in Victorian England* (1963).

Jackson, J. A., *The Irish in Britain* (1963).

Jackson Lears, T., 'The concept of cultural hegemony: problems and possibilities', *American Historical Review* 90:3 (1985), pp. 567–93.

Jaffray, J., *Hints for a history of Birmingham* (c. 1856).

Jenkins, M., *The general strike of 1842* (1980).
Jenkins Williams, D., *The Welsh church and Welsh people in Liverpool* (Liverpool, 1927).
Johnson, J., and Pooley, C. G., eds., *The structure of nineteenth-century cities* (1982).
Jones, D., ed., *Crime, protest, community and police in nineteenth-century Britain* (1982).
Jones, E., 'The Welsh in London in the seventeenth and eighteenth centuries', *The Welsh History Review* 10 (1981), pp. 461–79.
    'The Welsh in London in the nineteenth century', *Cambria* 12:1 (1985), pp. 149–69.
Jones, G. S., *Outcast London: a study in the relationship between classes in Victorian society* (Oxford, 1971).
    'Class expression versus social control? A critique of recent trends in the social history of "leisure"', *History Workshop* 4 (1977), pp. 162–70.
Jones, L. J., 'The public pursuit of private profit? Liberal businessmen and municipal politics in Birmingham, 1865–1900', *Business History* 25 (1983), pp. 240–59.
Joyce, P., *Work, society and politics: the culture of the factory in late Victorian England* (Brighton, 1980).
    ed., *The historical meaning of work* (Cambridge, 1987).
Kearns, G., *Urban epidemics and historical geography: cholera in London 1848–9* (Norwich, 1985).
    'Private property and public health reform in England 1830–70', *Social Science and Medicine* 26 (1988), pp. 187–99.
    'Death in the time of cholera', *Journal of Historical Geography* 15 (1989), pp. 425–32.
    'Zivilis or Hygaeia: urban public health and the epidemiologic transition' in Lawton, R., ed., *The rise and fall of great cities* (1989), pp. 96–124.
    'Cholera, nuisances and environmental management in Islington 1830–54', *Medical History* (in press).
Kearns, G., Lee, W. R., and Rogers, J., 'The interaction of political and economic factors in the management of urban public health' in Nelson, M. C., and Rogers, J., eds., *Urbanisation and the epidemiologic transition* (Uppsala, 1989), pp. 9–82.
Kirby, A. M., and Pinch, S. P., 'Territorial justice and service allocation' in Pacione, M., ed., *Progress in urban geography* (1983), pp. 223–50.
Kirk, N., 'Ethnicity, class and popular Toryism, 1850–1870' in Lunn, K., ed., *Hosts, immigrants and minorities: historical responses to newcomers in British society 1870–1914* (Folkestone, 1980), pp. 64–106.
Knox, P. L., 'Residential structure, facility location and patterns of accessibility' in Cox, K. R., and Johnston, R. J., eds., *Conflict, politics and the urban scene* (1982), pp. 62–87.
Kussmaul, A., *Servants in husbandry in early modern England* (Cambridge, 1981).
Lampard, E., 'The history of cities in the economically advanced areas', *Economic Development and Cultural Change* 3 (1954), pp. 81–136.
    'The urbanising world' in Dyos H. J., and Wolff, M., eds., *The Victorian city: images and realities*, vol. I (1973), pp. 3–57.
    'The nature of urbanisation' in Fraser, D., and Sutcliffe, A., eds., *The pursuit of urban history* (1983), pp. 3–53.
Langton, J., 'Liverpool and its hinterland in the late-eighteenth century' in Anderson,

B. L., and Stoney, P. J. M., eds., *Commerce, industry and transport: studies in economic change on Merseyside* (Liverpool, 1983), pp. 1–25.

'The industrial revolution and the regional geography of England', *Transactions, Institute of British Geographers* NS 9 (1984) pp. 145–67.

Langton, J., and Hoppe, G., *Town and country in the development of early modern Western Europe* (Norwich, 1983).

Large, D., 'The Irish in Bristol in 1851: a census enumeration' in Swift, R., and Gilley, S., eds., *The Irish in the Victorian city* (1985), pp. 37–58.

Law, C. M., 'The growth of urban population of England and Wales, 1801–1911', *Transactions, Institute of British Geographers* 41 (1967), pp. 125–43.

Lawton, R., 'Population trends in Lancashire and Cheshire from 1801', *Transactions of the Historic Society of Lancashire and Cheshire* 114 (1963), pp. 189–214.

ed., *The census and social structure: an interpretative guide to the nineteenth-century censuses for England and Wales* (1978).

Lecuyer, B. P., 'L' hygiène en France avant Pasteur, 1759–1850' in Salomon-Bayet, C., ed., *Pasteur et la révolution pastorienne* (Paris, 1986), pp. 65–139.

Lee, A., 'Aspects of the working-class response to the Jews in Britain, 1880–1914' in Lunn, K., ed., *Hosts, immigrants and minorities: historical responses to newcomers in British society 1870–1914* (Folkestone, 1980), pp. 107–33.

Lees, L. H., 'Patterns of lower-class life: Irish slum communities in nineteenth-century London' in Thernstrom, S., and Sennett, R., eds., *Nineteenth-century cities* (New York, 1976), pp. 359–85.

Exiles of Erin: Irish migrants in Victorian London (Manchester, 1979).

Leeson, R. A., *Travelling brothers* (1979).

Levine, D., and Wrightson, K., 'The social context of illegitimacy in early modern England' in Laslett, P., Oosterveen, K., and Smith, R., eds., *Bastardy and its comparative history* (1980), pp. 158–75.

Levitt, I., *Government and social conditions in Scotland, 1845–1919* (Edinburgh, 1988).

List, A. C. C., *The two phases of the social evil* (2nd edn, Edinburgh, 1861).

Lobban, R. D., 'The migration of Highlanders into Lowland Scotland c. 1750–1890 with particular reference to Greenock' (unpublished Ph.D. thesis, University of Edinburgh, 1969).

'The Irish community in Greenock in the nineteenth century', *Irish Geographer* 6 (1971), pp. 270–81.

Lojkine, J., 'Contribution to a Marxist theory of capitalist urbanisation' in Pickvance, C. G., ed., *Urban sociology: critical essays* (1976), pp. 119–46.

Lunn, K., ed., *Hosts, immigrants and minorities: historical responses to newcomers in British society 1870–1914* (Folkestone, 1980).

MacFarlane, A., *The culture of capitalism* (Oxford, 1987).

Macgregor, O. R., 'Social research and social policy in the nineteenth century', *British Journal of Sociology* 8: 2 (1957), pp. 146–57.

McKenzie, R., 'The ecological approach to the study of the human community' in Park, R. E., Burgess, E. W., and McKenzie, R., eds., *The city* (Chicago, 1925), pp. 63–79.

Maclaren, A. A., *Religion and social class: the disruption years in Aberdeen* (1974).

'Bourgeois hegemony and Victorian philanthropy: the contradictions of cholera' in Maclaren, A. A., ed., *Social class in Scotland: past and present* (Edinburgh, 1976), pp. 36–54.

ed., *Social class in Scotland: past and present* (Edinburgh, 1976).

McLennan, G., 'E. P. Thompson and the discipline of historical context' in Centre for Contemporary Cultural Studies/Johnson, R., et al., eds., *Making histories: studies in history-writing and politics* (1982), pp. 96–130.

McLeod, H., *Class and religion in the late-Victorian city* (1974).

'White collar values and the role of religion' in Crossick, G., ed., *The lower middle class in Britain* (1977), pp. 61–88.

Maehl, W. H., ed., *Robert Gammage: reminiscences of a Chartist* (Manchester, 1983).

Mahood, L., *The Magdalenes: prostitution in the nineteenth century* (1990).

Malchow, H. L., 'Public gardens and social action in late Victorian London', *Victorian studies* (1985), pp. 97–124.

Maldonado, L., and Moore, J., *Urban ethnicity in the United States* (1985).

Martin, B., *A sociology of contemporary cultural change* (Oxford, 1981).

Massey, D., *Spatial divisions of labour: social structures and the geography of production* (1984).

Matsumura, T., *The labour aristocracy revisited: the Victorian flint glass makers, 1850–80* (Manchester, 1983).

Matthison, A. L. L., *Less paint, more variety* (1937).

Mayhew, H., *London labour and the London poor. Volume I: The London street folk* (*Partial*) (New York, 1968; original edition, 1861–2).

London labour and the London poor. Volume II: The London street folk (*continued*) (New York, 1968; original edition 1861–2).

Meacham, S., *A life apart: the English working class, 1890–1914* (Cambridge, Mass., 1977).

Mechie, S., *The church and Scottish social development, 1780–1870* (1960).

Meller, H. E., *Leisure and the changing city, 1870–1914* (1976).

Mendelman, T., *The Jews of Georgian England* (Philadelphia, Penn., 1979).

Millward, P., 'The Stockport riots of 1852: a study of anti-Catholic and anti-Irish sentiment' in Swift, R., and Gilley, S., eds., *The Irish in the Victorian city* (1985), pp. 207–24.

Molyneux, D. D., 'The development of physical recreation in the Birmingham district from 1871 to 1892' (unpublished MA thesis, University of Birmingham, 1957).

Morris, R. J., *Cholera 1832: the social response to an epidemic* (1976).

'Voluntary societies and British urban élites, 1780–1850: an analysis', *Historical Journal* 26 (1983), pp. 95–118.

ed., *Class, power and social structure in nineteenth-century British cities* (Leicester, 1987).

Mouffé, C., ed., *Gramsci and Marxist theory* (1979).

Muirhead, I. A., 'Churchmen and the problem of prostitution in nineteenth century Scotland', *Records of the Scottish Church History Society* 18 (1972–4).

Murray, C., *Losing ground* (New York, 1984).

Musson, A. E., *The Typographical Association* (Oxford, 1954).

Neale, R. S., 'Cultural materialism', *Social History* 9:2 (1984), pp. 199–215.

Newman, O., *Defensible space: people and design in the violent city* (1972).

Nicol, J., *A voice from the bothie* (Aberdeen, 1851).

Nield, K., and Seed, J., 'Waiting for Gramsci', *Social History* 6:2 (1981), pp. 209–27.

Noble, A., 'Urbane silence: Scottish writing and the nineteenth century city' in Gordon, G., ed., *Perspectives on the Scottish city* (Aberdeen, 1985), pp. 64–90.

Olsen, D. J., *The growth of Victorian London* (1976).

O'Tuathaigh, M. A. G., 'The Irish in nineteenth-century Britain: problems of integration' in Swift, R., and Gilley, S., *The Irish in the Victorian city* (1985), pp. 13–36.

Palfry, J., *Mission work among the destitute, or scenes in the abodes of the poor* (1867).

Park, R. E., 'Human migration and the marginal man' in Sennett, R. ed., *Classic essays on the culture of cities* (New York, 1969), pp. 131–42.

'The crowd and the public' in Elsner, H., ed., *Robert E. Park: the crowd and the public and other essays* (Chicago, Ill., 1972).

Park, R. E., and Burgess, E. W., eds., *Introduction to the science of sociology* (Chicago, Ill., 1921).

Parker, R. D., 'The changing character of Preston Guild Merchants, 1762–1862', *Northern History* 20 (1984), pp. 108–26.

Peach, C., ed., *Urban social segregation* (1975).

Pearson, G., *Hooligan: a history of respectable fears* (1983).

Phillips, A., *A history of the origins of the first Jewish community in Scotland – Edinburgh 1816* (Edinburgh, 1979).

'The physical condition of the people: the agricultural labourer', *Journal of Agriculture* 21 (1857–9), pp. 623–41.

Pinch, S. P., *Cities and services: the geography of collective consumption* (1985).

Poole, R., 'Oldham wakes' in Walton, J. K., and Walvin, J., eds., *Leisure in Britain* (Manchester, 1983), pp. 72–98.

Pooley, C. G., 'The residential segregation of immigrant communities in mid-Victorian Liverpool', *Transactions, Institute of British Geographers* NS 2 (1977), pp. 364–82.

'Welsh migration to England in the mid-nineteenth century', *Journal of Historical Geography* 9:3 (1983), pp. 287–306.

'Book review: Dennis, "English industrial cities"', *Progress in Human Geography* 10 (1986), p. 615.

Poulantzas, N., 'Marxist political theory in Great Britain', *New Left Review* 43 (1967), pp. 18–43.

Pred, A., *City-systems in advanced economies: past growth, present processes and future development options* (1977).

Price, R., *Masters, unions, and men: work control building and the rise of labour 1830–1914* (Cambridge, 1980).

Pringle, R. H., 'The agricultural labourer of Scotland – then and now', *Transactions of the Highland and Agricultural Society of Scotland*, 5th series, 6 (1894), pp. 238–71.

Pritchard, R. M., *Housing and the spatial structure of the city* (Cambridge, 1976).

Prothero, I., *Artisans and politics in early nineteenth century London: John Gast and his times* (Baton Rouge, 1979).

Pryce, W. T. R., 'Migration and the evolution of culture areas: cultural and linguistic frontiers in north-east Wales, 1750–1851', *Transactions, Institute of British Geographers* 65 (1975), pp. 79–107.

Randall, A. J., 'The shearmen and the Wiltshire outrages of 1802: trade unionism and industrial violence', *Social History* 7 (1982), pp. 283–304.

Reid, D. A., 'Interpreting the festival calendar: wakes and fairs as carnivals' in Storch, R. D., ed., *Popular culture and custom in nineteenth-century England* (1982), pp. 125–53.

'Labour, leisure and politics in Birmingham, c. 1800–1875' (unpublished Ph.D. thesis, University of Birmingham, 1985).

Reissman, L., *The urban process: cities in industrial societies* (New York, 1964).

Report by a committee of the prison board of Aberdeenshire on the repression of prostitution (Aberdeen, n.d. [1860]).

Report of the Committees of Management of the Auxiliary Society of Glasgow for the Support of Gaelic Schools (Glasgow, 1815).

Richardson, C., 'Irish settlement in mid-nineteenth century Bradford', *Yorkshire Bulletin for Economic and Social Research* 20 (1968), pp. 40–57.

Roberts, E., *A woman's place: an oral history of working-class women, 1890–1940* (Oxford, 1984).

Robson, B., *Urban analysis: a study of city structure with special reference to Sunderland* (Cambridge, 1969).

Robson, B. T., *Urban growth: an approach* (1973).

Rodger, R. G., 'The building cycle and the urban fringe in Victorian cities: another comment', *Journal of Historical Geography* 5 (1979), pp. 72–8.

Rose, G., 'Locality-studies and waged labour: an historical critique', *Transactions, Institute of British Geographers* NS 14 (1989), pp. 317–28.

Rosenberg, C., *The cholera years: the United States in 1832, 1849 and 1866* (Chicago, Ill., 1962).

Ross, E., 'Survival networks: women's neighbourhood sharing in London before World War One', *History Workshop Journal* 15 (1983), pp. 4–27.

Roth, G., *The rise of provincial Jewry* (1950).

de Rousier, P., *The labour question in Britain* (1896).

Roy Lewis, C., 'The Irish in Cardiff in the mid-nineteenth century', *Cambria* 7 (1980), pp. 13–41.

Rule, J., *The experience of labour in eighteenth-century industry* (1981).

Ryan, W., *Blaming the victim* (New York, 1976).

Samuel, R., 'Comers and goers', in Dyos, H. J. and Wolff, M., eds., *The Victorian city: images and realities*, vol. I (1973), pp. 123–60.

Sartre, J. P., *Critique of dialectical reason* (1976).

Saunders, P., *Social theory and the urban question* (1981).

Schneider, L., and Bonjean, C., eds., *The idea of culture in the social sciences* (Cambridge, 1973).

Schnore, L., 'Geography and human ecology', *Economic Geography* 37 (1961), pp. 207–17.

Schwarzbach, F. S., *Dickens and the city* (1979).

Scott, J. W., and Tilly, L. A., 'Women's work and the family in nineteenth-century Europe', *Comparative Studies in Society and History* 17:1 (1975), pp. 36–64.

'Senex' [R. Reid], *Glasgow past and present* (Glasgow, 1884).

Sennett, R. ed., *Classic essays on the culture of cities* (New York, 1969).

The fall of public man (Cambridge, 1977).

Seton, G., *The causes of illegitimacy particularly in Scotland* (Edinburgh, 1860).

'Shadow', *Midnight scenes and social photographs: being sketches of life in the streets, wynds and dens of the city* (Glasgow, 1858).

Sharpless, J., and Lindstrom, K., 'Urban growth and economic structure in Antebellum America', *Research in Economic History* 8 (1978), pp. 161–216.

Shaw, M., 'The ecology of social change: Wolverhampton, 1851–71', *Transactions, Institute of British Geographers* NS, 2 (1977), pp. 332–48.

Shevky, E., and Bell, W., *Social area analysis: theory, illustrative application and computational procedures* (Westport, Conn., 1955).

Shorter, E., 'Illegitimacy, sexual revolution and social change in modern Europe' in Rotberg, R. I., and Rabb, T. K., eds., *Marriage and fertility: studies in interdisciplinary history* (Princeton, N. J., 1980), pp. 85–120.

Showstack Sassoon, A., *Gramsci's politics* (1980).

Sider, G., 'The ties that bind: culture and agriculture, property and propriety in the Newfoundland village fishery', *Social History* 5 (1980), pp. 1–39.

Smail, J., 'New language for labour and capital: the transformation of discourse in the early years of the Industrial Revolution', *Social History* 12:1 (1987), pp. 49–71.

Smith, D., *Conflict and compromise: class formation in English society, 1830–1914: a comparative study of Birmingham and Sheffield* (1982).

Smith, J., 'Class, skill and sectarianism in Glasgow and Liverpool, 1880–1914' in Morris, R. J., ed., *Class, power and social structure in nineteenth-century British cities* (Leicester, 1987), pp. 184–215.

Smith, M. P., *The city and social theory* (Oxford, 1980).

Smout, T. C., 'Aspects of sexual behaviour in nineteenth century Scotland', in Maclaren, A. A., ed., *Social class in Scotland: past and present* (Edinburgh, 1976), pp. 55–85.

Sontag, S., *AIDS and its metaphors* (1989).

Southall, H. R., 'Regional unemployment patterns in Britain, 1851 to 1914: a study of the trade union percentages of unemployment, with special reference to engineering workers' (unpublished Ph.D. thesis, University of Cambridge, 1984).

'Towards a geography of unionization: the spatial organization and distribution of early British trade unions', *Transactions, Institute of British Geographers* NS 13 (1988) pp. 466–83.

'The tramping artisan revisits: labour mobility and economic distress in early Victorian England', *Economic History Review*, 2nd series, 44 (1991, in press).

Springett, J., 'Landowners and urban development: the Ramsden Estate and nineteenth-century Huddersfield', *Journal of Historical Geography* 8 (1982), pp. 129–44.

Stallybrass, P., and White, A., *The politics and poetics of transgression* (1986).

Stephens, W. B., ed., *The city of Birmingham, Victoria county history of Warwickshire, Volume VII* (Oxford, 1964).

Stephenson, S. J., 'The "criminal class" in the mid-Victorian city: a study of policy in Birmingham and East London' (unpublished Ph.D. thesis, University of Oxford, 1983).

Storch, R. D., 'The policeman as domestic missionary: urban discipline and popular culture in northern England, 1850–80', *Journal of Social History* 9 (1976), pp. 481–509.

Strachan, A. J., and Bowler, I. R., 'The development of public parks and gardens in the city of Leicester', *East Midlands Geographer* 46 (1976), pp. 275–83.

Strachan, J., *Address upon illegitimacy to the working men of Scotland* (Edinburgh, n.d. [1859]).

Strachan, J. M., 'Immorality in Scotland' *The Scotsman*, 20 May 1870.

Stuart, H., *Agricultural labourers as they were, are, and should be in their social condition* (Edinburgh, 1853).

Swenarton, M., 'Notes on the concept of urban space' (unpublished paper, Urban History Group meeting, University of London, 1984).

Swift, R., and Gilley, S., eds., *The Irish in the Victorian city* (1985).

Tait, W., *Magdalenism: an inquiry into the extent, causes and consequences of prostitution in Edinburgh* (2nd edn, 1842).

Teitelbaum, M. S., *The British fertility decline: demographic transition in the crucible of the Industrial Revolution* (Princeton, N. J., 1984).

Thernstrom, S. A., *The other Bostonians: poverty and progress in the American metropolis, 1880–1970* (Cambridge, Mass., 1973).

Thomas, K., 'The double standard', *Journal of the History of Ideas* 20:2 (1959), pp. 195–216.

Man and the natural world: changing attitudes in England, 1500–1800 (1983).

Thompson, D., *The Chartists* (Aldershot, 1984).

Thompson, E. P., *The making of the English working class* (Harmondsworth, Middx., 1968; original edition 1963).

'The moral economy of the English crowd in the eighteenth century', *Past and Present* 50 (1971), pp. 76–136.

'Patrician society, plebeian culture', *Journal of Social History* 6 (1974), pp. 382–405.

The poverty of theory (1974).

Thompson, F. M. L., 'Introduction' in Thompson, F. M. L., ed., *The rise of suburbia* (Leicester, 1982), pp. 2–25.

Thomson, A., *Social evils: their causes and their cure* (1852).

'A brief history of preventative and reformatory work in Aberdeen', *Transactions of the National Association for the Promotion of Social Science* (1860), pp. 513–25.

Thorne, W., *My life's battles* (1925).

Thrift, N., and Williams, P., eds., *Class and space: the making of urban society* (1987).

Thurley, E. C., 'Satan and Sambo: the image of the immigrant in English racial populist thought since the First World War' in Lunn, E., ed., *Hosts, immigrants and minorities: historical responses to newcomers in British society 1870–1914* (Folkestone, 1980), pp. 39–63.

Turner, V. W., *The ritual process: structure and anti-structure* (1969).

Urry, J., 'Society, space and locality', *Environment and Planning, D, Society and Space* 5 (1987), pp. 435–44.

Vacher, F., *Seduction in Edinburgh: being some particulars regarding the relative social position of the seducer and the seduced* (Edinburgh, 1867).

Valentine, J., 'Illegitimacy in Aberdeen and other large towns of Scotland', *Report of the British Association for the Advancement of Science* (1859), pp. 224–6.

Vance, J. E., 'Housing the worker: the employment linkage as a force in urban structure', *Economic Geography* 42 (1966), pp. 294–325.

'Housing the worker: determinant and contingent ties in nineteenth-century Birmingham', *Economic Geography* 43 (1967), pp. 95–127.

This scene of man: the role and structure of the city in the geography of western civilisation (New York, 1977).

de Vries, J., *European urbanisation 1500–1800* (1984).

Walton, J. K., 'The Lancashire wakes in the nineteenth century' in Storch, R. D., ed., *Popular culture and custom in nineteenth-century England* (1982), pp. 100–24.

Walvin, J., *A child's world: a social history of English childhood, 1800–1914* (Harmondsworth, Middx., 1982).

Ward, D., 'The pre-urban cadastre and the urban pattern of Leeds', *Annals, Association of American Geographers* 52 (1962), pp. 150–66.

Cities and immigrants (New York, 1970).

'Victorian cities: how modern?', *Journal of Historical Geography* 1 (1975), pp. 135–51.

'The Victorian slum: an enduring myth?', *Annals, Association of American Geographers* 66 (1976), pp. 323–36.

'Environs and neighbours in the "Two Nations": residential differentation in mid-nineteenth century Leeds', *Journal of Historical Geography* 6 (1980), pp. 133–62.

'The ethnic ghetto in the United States: past and present', *Transactions, Institute of British Geographers* NS 7 (1982), pp. 257–75.

Poverty, ethnicity and the American city, 1840–1925: changing conceptions of the slum and the ghetto (Cambridge, 1989).

Warner, S. B., *Streetcar suburbs: the process of growth in Boston, 1870–1900* (Cambridge, Mass., 1962).

Watson, W., 'The Poor Law from the poor man's standpoint' in Ivory, T., ed., *Pauperism and the Poor Laws* (Edinburgh, 1870), 24 pp. separately paginated.

Pauperism, vagrancy, crime and industrial education in Aberdeenshire 1840–1875 (Aberdeen, 1877).

'Rural police in Scotland', *Transactions of the National Association for the Promotion of Social Science* (1877), pp. 314–55.

Webb, I., *From custom to capital: the English novel and the Industrial Revolution* (Ithaca, N.Y., 1981).

Webb, S., and Webb, B., *The history of trade unionism* (1894).

Weeks, J., *Sex, politics and society: the regulation of sexuality since 1800* (1981).

Weinberger, B., 'Law breakers and law enforcers in the late-Victorian city: Birmingham, 1867–1877' (unpublished Ph.D. thesis, University of Warwick, 1981).

'The police and the public in mid-nineteenth-century Warwickshire' in Bailey, V., ed., *Policing and punishment in nineteenth-century Britain* (1981).

A well meant word to the women of Aberdeenshire and other north-eastern counties (Aberdeen, n.d.).

Wells, R., *Insurrection: the British experience 1795–1803* (Gloucester, 1983).

Werly, J. M., 'The Irish in Manchester, 1832–1849', *Irish Historical Studies* 18 (1973), pp. 345–58.

White, J., *Rothschild buildings: life in an East End tenement block, 1887–1920* (1980).

The worst street in North London: Campbell Bunk, Islington, between the wars (1986).

White, L., 'The concept of culture', *American Anthropologist* 61 (1959), pp. 229–37.

White, W., *Life in a court: how to stop neighbours' quarrels* (Birmingham, 1890), p. 3.

Whitehand, J. W. R., 'Building activity and intensity of development at the urban fringe: the case of a London suburb in the nineteenth century', *Journal of Historical Geography* 1 (1975), pp. 211–24.

'The building cycle and the urban fringe in Victorian cities: a reply', *Journal of Historical Geography* 4 (1978), pp. 181–91.

Wiener, M., *English culture and the decline of the industrial spirit 1850–1980* (Cambridge, 1981).

Wilkinson, D., *Rough roads: reminiscences of a wasted life* (1912).

Williams, A. S., *The rich man and the diseased poor in early-Victorian literature* (1987).

Williams, B., *The making of Manchester Jewry* (Manchester, 1976).

'The beginnings of Jewish trade unionism in Manchester, 1889–1891' in Lunn, K., ed., *Hosts, immigrants and minorities: historical responses to newcomers in British society 1870–1914* (Folkestone, 1980), pp. 263–307.

Williams, C. H., ed., *Linguistic minorities, society and territory* (Clevedon, 1991).

Williams, G. A., 'The concept of *egomonia* in the thought of A. Gramsci', *Journal of the History of Ideas* 21 (1960), pp. 468–99.

Williams, J. G., 'The impact of the Irish on British labour markets during the Industrial Revolution', *Journal of Economic History* 46 (1986), pp. 693–720.

Williams, R., 'Base and superstructure in Marxist cultural theory', *New Left Review* 82 (1973), pp. 3–16.

The country and the city (St Albans, 1975).

Marxism and literature (Oxford, 1977).

Wilson, J. D., 'Law of infanticide. Can any better measures be devised for the prevention and punishment of infanticide?', *Transactions of the National Association for the Promotion of Social Science* (1877), pp. 284–94.

Withers, C. W. J., *Gaelic in Scotland 1698–1981: the geographical history of a language* (Edinburgh, 1984).

'Highland clubs and Gaelic chapels: Glasgow's Gaelic community in the eighteenth century', *Scottish Geographical Magazine* 101:1 (1985), pp. 16–27.

'Highland migration to Dundee, Perth and Stirling, 1753–1891', *Journal of Historical Geography* 11:3 (1985), pp. 395–418.

'Kirk, club and culture change: Gaelic chapels, Highland societies and the urban Gaelic sub-culture in eighteenth-century Scotland', *Social History* 10:2 (1985), pp. 171–92.

Highland communities in Dundee and Perth, 1787–1898 (Dundee, 1986).

'Poor relief in Scotland and the General Register of Poor', *Local Historian* 17:1 (1986), pp. 19–29.

'Destitution and migration: labour mobility and relief from famine in Highland Scotland 1836–1850', *Journal of Historical Geography* 14:2 (1988), pp. 128–50.

Gaelic Scotland: the transformation of a culture region (1988).

'On the geography and social history of Gaelic' in Gillies, W., ed., *Gaelic and Scotland* (Edinburgh, 1989), pp. 101–30.

'Historical urban geolinguistics: Gaelic speaking in urban Lowland Scotland in

1891' in Williams, C. H., ed., *Linguistic minorities, society and territory* (Clevedon, 1991), pp. 212–40.

'Gaelic speaking in urban Lowland Scotland: the evidence of the 1891 Census', *Scottish Gaelic Studies* 16:1 (1991), pp. 103–39.

Wolff, K. H., ed., *The sociology of Georg Simmel* (New York, 1950).

Woods, D., 'Community violence' in Benson, J., ed., *The working class in England, 1875–1914* (1985), pp. 165–205.

Woods, R., 'Mortality and sanitary conditions in the "Best governed city in the world" – Birmingham, 1870–1910', *Journal of Historical Geography* 4 (1978), pp. 35–56.

Wrigley, E. A., 'A simple model of London's importance in changing English society and economy, 1650–1750', *Past and Present* 37 (1967), pp. 44–70.

ed., *Nineteenth-century society: essays in the use of quantitative methods for the study of social data* (Cambridge, 1972).

Yancey, W. L., Ericksen, E. P., and Juliani, R. N., 'Emergent ethnicity: a review and reformulation', *American Sociological Review* 41 (1976), pp. 391–403.

Zohn, H., ed., *Marianne Weber – Max Weber: a biography* (1975).

Zunz, O., *The changing face of inequality: urbanisation, industrial development and immigration in Detroit, 1880–1920* (Chicago, Ill., 1982).

Index

Aberdeen, 62, 63, 65, 66, 87, 90, 92, 94, 96, 99, 101
Aberdeen, Countess of, 95, 97
Aberdeenshire, 86, 93
   Prison Board for, 92
   Sheriff-Substitute for, 92
Adams, William, 113, 127, 128
affinity groups, 104, 119
agitators, travelling, 113, 126–7
Agnew, John, 61
agricultural depression, 89, 94
AIDS, 89
Alison, William, 13
Allan, William, 123
Amsterdam, 91
Anderson, Perry, 129
Angus and Mearns, Synod of, 86
Anti-Corn Law League, 127
Applegarth, Robert, 123
Argyllshire, 63, 70, 71
Armstrong, William, 14, 15
Arnott, Neil, 13
artisans, 11, 15, 88, 99, 103, 105, 106, 107, 113, 118, 122, 123, 128, 129

bailiffs, 40
Balzac, Honoré, 21
Banff, 88
Banffshire, 83, 84, 86, 88, 93, 94, 98, 99
bastardy, 82, 83, 85, 87, 90, 92, 97, 98, 101
bastardy ratios, 83, 84, 93, 97
baths, 49
Begg, Rev. James, 81, 83, 86, 87, 101
behaviour, riotous, 10, 36, 44, 45, 49, 52, 53
Benbow, William, 126

Beresford, Maurice, 3
Berlin, 91
Berry, Brian, 6
Billinge, Mark, 61
Birmingham, 33–54 passim, 109
Blackburn, 109
Booth, Charles, 20
bothies, 83, 86
Boyd, Kenneth, 87
Bradford, 56, 112
bricklayers, 118
Brief Institution, 104
Bristol, 56, 109
Broadhurst, Henry, 116, 117, 119
Brookfield, Harold, 4

cabinet makers, 15, 107
Cannadine, David, 4
capitalism, 4, 31, 54, 85, 89, 129
Cardiff, 56
Carpenters and Joiners, Amalgamated Society of, 123
Carter, Ian, 85
Cartwright, Major, 123–5
Celts, ethnological attitudes to, 57, 67–8, 77
census, Gaelic, 64–79 passim
Chadwick, Edwin, 9, 12, 13, 22, 23, 24, 28, 29, 90
Chalmers, Thomas, 82
chapels, Gaelic, 56–7, 62–77 passim
Chartism, 26, 84, 106, 107, 109, 112–3, 123–7
chastity, 100
Checkland, Olive, 99
Cheltenham, 113, 117, 127

172

Cheshire, 126
Chevalier, Louis, 21, 22, 23, 26
Chicago school, 2, 3
Chinese, 57
cholera, 18, 19, 21, 90, 94
civil registration (Scotland), 81
clans, 59, 60, 71, 76
class
   as a set of relations, 7–9, 11, 32–3, 55,
      57, 59, 73, 84, 127–9
   as biology, 9, 12–13, 14, 21–30 *passim*
   as class consciousness, 8, 32, 57–8, 97
   metaphors of, 9, 11
class formation, 7, 8, 57, 129
Coachmakers, United Kingdom Society of,
   109
coach trimmers, 109, 112
Cobbett, William, 9
Cohen, Stanley, 97
Coleman, Brian, 22
Combination Acts, 105
community, 2, 6, 9, 31–54 *passim*, 56, 58,
   73, 82, 100, 103, 122, 128–9
   civic, 52, 53
   working class, 31–52 *passim*, 127–9
companies, London, 104
consciousness, contradictory, 8, 32, 55, 57,
   59, 62, 67, 74, 122
contagion, 12, 23, 27, 89, 90
cottars, 82, 85
courtship, 87, 88, 101
courtyards, 32, 33, 36, 38, 42
craft traditions, 10, 22, 118, 122
Cramond, William, 88
crime, 35, 51
Crossick, Geoffrey, 3, 15
'cultural coherence', idea of, 58, 98
'cultural distance', idea of, 58, 98
culture, 6, 8, 10, 55–79 *passim*, 106, 127–9
   'emergent', 60
   'residual', 10, 60, 62
   working class, 10, 31–52 *passim*, 127–8

Darby, Sir H.C., 1, 2
Darwinism, social, 2, 95
Daunton, Martin, 34–5, 36, 37, 39, 45, 48,
   53
Davidoff, Leonore, 34
Dennis, Richard, 1, 5, 79
deprivation, transmitted, 82, 93
Derby, 109
dialect, 70, 117

Dickens, Charles, 22, 23, 24, 26, 27
disease, 2, 4, 12, 16, 22, 23, 26, 27, 28, 90,
   92
   sexually transmitted, 26–7, 89
   venereal, attitudes to, 89
drink, 52, 107
Duke Street chapel, Glasgow, 64, 66, 69,
   70, 74
Dundee, 56, 62, 63, 65, 66, 82, 92, 127
Dupré, Leonard, 60
Dyos, Harold, 3

ecological anthropology, 25, 28, 59, 67–8
economy
   moral, 3–4, 99
   political, 3–4, 99
Edinburgh, 56, 62, 63, 82, 86, 90, 92, 112,
   118, 125, 126
emigration, 65, 90, 116
engineers and mechanics, 107, 118, 122
Engineers, Amalgamated Society of, 123
English language, cultural authority of, 62,
   67, 70, 73, 76, 77, 78
environmentalism, 21, 27, 82, 97
equalisation of funds, 105
ethnicity, 58, 73, *see also* Chinese,
   Highlanders, Irish, Jews, migrant
   identity, Welsh
excretion, moral attitudes towards, 13, 29

factory workers, 91, 93, 99, 103
Fairbairn, William, 116
family reconstitution, 85, 89
Farr, William, 23
fertility, 21, 83, 87, 89
   extra-marital, 83, 87
   index of, 83
   testing, 87, 89
filth, attitudes to, 13, 29
fishing villages, 83, 84
fluids, bodily, 29
footings, 117
Foster, John, 3
Free Church of Scotland, 83, 86, 101
friendly societies, 104, 105, 113, 120
Friendly Societies, Registrar of, 105

Gaelic Club of Gentlemen, 76
Gaelic language, 56, 64–5
Gaelic Schools, Glasgow Auxiliary Society
   for the Support of, 75–6
gambling, 44, 46, 49

Gammage, Robert, 109–13, 117, 127, 128
gas workers, 121
Gast, John, 128
Gay, Peter, 29
ghettos, 9, 56
Gilbert, E.W., 2
Glasgow, 62–78 *passim*, 82, 92, 112, 121,
    123, 125, 126
Glasgow Argyllshire Society, 71
Glasgow Celtic Infirmary, 76
Glasgow Highland Society, 67, 75–6
gospel, civic, 52
gossip, 40
Gramsci, Antonio, 59
Gray, Malcolm, 80, 85
Gray, Robert, 3, 61
Greenock, 56, 62, 63
Gregory, Derek, viii, 4

Habermas, Jürgen, 4
Haddo House, 95, 96
Hall, Catherine, 34
Hamburg, 91
Harney, George Julian, 112
Harvey, David, 3, 4, 80
Health, Boards of, 14–18
'health divide', 14, 17–19, 23, 48, 80
hegemony, 7, 10, 56, 58–64 *passim*, 74–9
    passim, 84
  alternative, 60, 74
  counter, 60, 61, 74, 78
Highlanders, 55, 62, 64, 65, 67
Hobsbawm, Eric, 3, 103, 104
Hope Street chapel, Glasgow, 64
houses of call, 105, 109–10, 112, 120–2
housing, 2, 80, 83, 86, 100
Housing, Free Church of Scotland
    Committee on, 84, 86, 101
Howe, Melvyn, 2
Huddersfield, 56
Hugo, Victor, 21
Hull, 108, 109, 123, 126
Hunt, Henry, 125
hygiene, 21–2, 29, 92

ideology, 6, 21–30 *passim*, 78, 81, 84, 96,
    97, 98, 101
illegitimacy, 82, 85, 87, 90, 91, 92, 93–4,
    97, 99, 100
  ratios of, 82, 91, 92
  returns on, 82, 88, 91, 97
illiteracy, 83

immorality, 83, 87, 90, 97, 98, 101
indoor relief, 21
industrialisation, 1, 3, 100
infanticide, 92
Ingram Street chapel, Glasgow, 62, 67, 69,
    71
intellectuals, 59
intemperance, 83, 91
Inverness-shire, 63, 70
Irish, 42, 43, 56, 57, 58, 62, 68, 70, 73, 77,
    89
Irish poor, Report on the state of the, 56

Jews, 25, 34, 54, 77
Johnson, James, 2
juveniles, life of, 44, 46, 47

Kay, J.P., 56
Keith, 88
kin networks, as source of assistance to
    migrants, 57, 70, 74, 119
kinship, as source of assistance to
    travelling artisans, 69–70, 110,
    119

labour
  aristocracy, 3
  division of, 8, 58
  market, 8, 20, 56, 77, 99, 114
  process, 56, 100, 103
Lampard, Eric, 21
Lancashire, 107, 112, 126
Langton, John, 129, 130
language, cultural identity from use of, 57,
    59, 60, 65
Lawton, Richard, 2
Lécuyer, B., 29
Leeds, 3, 8, 56, 109, 112
Leeson, Robert, 104, 105
Lenin, Vladimir, 3
Life and Work, Church of Scotland
    Committee on, 84
life-cycle stage, 14, 85, 115–16, 117, 122
Liverpool, 56, 58, 109, 123, 126
locality, ideas of community expression
    and, 11, 32, 35, 39, 47, 70, 129
lodging houses, 13, 20, 120
London, 13, 15, 17, 18, 19, 20, 22, 25, 26,
    28, 34, 36, 56, 57, 91, 109, 112, 118,
    119, 123, 126, 127, 129
Lothians, 86, 89
Lowery, Robert, 126

Maclaren, Allan, 80, 90
Macleod, Rev. Norman, 64, 70, 74, 76, 78
Malchow, H.L., 36
Malthus, Rev. Thomas, 86
Malthusianism, 25
Manchester, 56, 61, 108, 125, 126
marginal man, theory of migrant as, 58
markets, hiring and feeling, 83, 93, 100
marriage, 9, 82, 86, 89, 100, 118–19, 122, 123
marriage ratio, 89, 99
Marx, Karl, 7
Marxism, 3, 4, 54
masturbation, 29
Mayhew, Henry, 13, 22, 23, 25, 28, 30
Meacham, Stanley, 35, 53
Middlesex, 107
migrant identity, 8, 55–79 *passim*
migrants, 7, 9, 55–64, *see also* Chinese, Highlanders, Irish, Jews, Welsh
migration, 1, 2, 6, 21, 35, 82–3, 85, 90, 94, 97, 103, 104
    return, 114
    seasonal, 1, 6, 121
miners, 121, 123, 129
Mitchell, Joseph, 126
mobility
    occupational, 7, 13, 20, 85, 100
    residential, 71–2, 73, 85, 103, 106
moral topography, 9, 12, 29, 81, 83, 97
mortality
    age-specific, 13, 17, 18, 24
    maternal, 97
    occupational, 13, 14, 15, 17, 18, 20, 24
    peri-natal, 91
    sex differentials in, 14, 15, 17, 18, 24
mothers, unmarried, 99
Mouffé, Chantal, 61

neighbourhoods, 31, 32, 34, 37, 39, 40, 41, 43, 44, 49, 52, 53
Newcastle, 109, 112, 125, 126
Newcastle and County United Tanners, 114
Newman, Oscar, 38
newspapers, 48, 51, 52
Newton, William, 123
Noble, Andrew, 101
Northern Star, 112, 113, 126

O'Connor, Feargus, 124–6 *passim*
Old Machar parish, 93

Olsen, Donald, 34
Onward and Upward (Association), 95
Onward and Upward (magazine), 95, 96, 97, 99, 100
oral history, 35, 106, 107
order, public, 2, 10, 35, 36
outdoor relief, 13, 93
overcrowding, 20

Paisley, 62, 63, 125
Palermo, 91
Parent-Duchâtelet, J.P.B., 29
Paris, 21, 22, 80
Park, Robert, 2, 3
parks, municipal, 32, 33, 34, 36, 42, 48, 49, 50
pauperism, 12, 71, 82, 94
paupers, 12, 19, 21, 71, 82, 85, 94
'peaky blinders', 47
Perth, 56, 63, 65, 66
philanthropism, 75, 76, 78, 85, 86, 92–97 *passim*, 100
phthisis, 16
Pilling, Richard, 126
police, 35, 42, 43, 44, 45, 85, 93, 96, 100
Pooley, Colin, 2, 5
Poor Law, 13, 19, 20, 76, 85, 93, 100
Poor Law Amendment Act (Scotland), 81, 85, 93
Poor Law inspectors, 20, 94
poorhouses, 13, 20, 94
population density, 91
positivism, 2, 4
Poulantzas, Nicos, 61
poverty, 4, 12, 13, 74, 91, 92
pregnancy, bridal, 87, 89
Presbyterianism, 78, 100
printers and compositors, 105, 107, 121, 122, 126, 128
property, private, 12, 80, 87
prostitution, 26, 29, 82, 83, 87, 89, 90, 91, 93, 95, 96, 100, 101
public health movement, 12, 22, 23, 26, 90
public houses, 44, 104, 105, 120

racial theories, 56, 57, 67–8, 83, 94
railways, 14, 105, 118
Ravenstein, Ernst, 2
Rechabites Friendly Society, 120
recreation, 32, 46, 49, 51, 54
reformatories, 90

Registrar General (Scotland), 79, 82, 83, 86, 92, 97
Registration Examiner (Scotland), 84, 96
Reid, Douglas, 44, 53
religion, 54, 57, 66, 68, 83, 107
Religion and Morals, Free Church of Scotland Committee on, 84, 86
Religious Condition of the People, Church of Scotland Commission on, 84
Religious Instruction, Commissioners of, 56, 64
rent, 60
respectability, 53, 54, 88, 90, 120
riots, 31, 33, 35
Robson, Bryan, 3
Romanticism, 77, 82
Rose, Rev. Lewis, 64, 76
Ross-shire, 63, 70

St Columba's parish, Glasgow, 64, 66
St George's parish, Glasgow, 64
St Nicholas' parish, Aberdeen, 93
Sanitary Condition of the Labouring Population, Report on, 24
sanitary idea, 12, 23–5
Sauer, Carl, 1
schools, 38, 41, 47, 48, 76, 93
Science, British Association for the Advancement of, 94
science, spatial, 2
Scotland, 55–78 *passim,* 80–102 *passim,* 123
seat rents, 68–9
Seditious Societies Act, 123
seduction, 89–90
segregation
  residential, 1, 31, 57, 70, 73, 74
  social, 2, 8, 31, 70, 73, 74, 77
servants
  domestic, 20, 90, 91, 94, 95, 96
  farm, 83, 84, 85, 86, 95, 97
  life-cycle of, 83, 88
sewers, 12, 25, 28, 29, 30
sex
  oral, 29
  ratios, 83
sexuality, ideas on, 27–9, 80, 89, 92, 100
Sheffield, 112, 123
shoemakers, 113, 117, 121, 126, 127
Simmel, Georg, 2, 3
skill, as determinant of mobility, 107, 114, 127, 129

slums, 9, 12, 27, 30, 39, 42, 43, 94
Smith, David, 75
Smout, Christopher, 89, 100
socialism, 107, 126
Southampton, 109
Southwood Smith, Thomas, 12
space
  defensible, 38, 39
  public, 10, 31–54 *passim*
sports, cruel, 44, 45
Stallybrass, P., 29
status, 39, 57
Steam Engine Makers' Society (SEM), 107–10, 114–6, 122
Stedman Jones, Gareth, 4
Stirling, 63, 65, 66
stonemasons, 107, 118, 119
Strathbogie, Presbytery of, 85
street
  gangs, 46
  life, 10, 32, 33, 35, 40, 45, 46, 47, 48, 53
syphilis, 89

tailors, 105, 126
tanners, 114, 118
temperance, 83, 91, 107, 126
Thompson, Dorothy, 106
Thompson, Edward, 3, 7, 60, 123, 129
Thompson, Michael, 34
tools, importance to travellers of, 120, 122
trade cycle, relation of migration to, 114, 116, 119–20
trade unions,
  as source of assistance to travellers, 112, 119–20
  over-emphasis by labour historians on, 103
  role of travelling in development of, 106, 123, 128
  types of benefit, 104–5
tramps, *see* travel
travel
  by foot, 117–8
  by rail, 105, 118, 121, 125
  by road vehicle, 118, 120, 121, 125
  by sea, 108–9, 118
typhus, 18, 24

Uist and Barra Association, 71
underclass, 10, 92
unemployment, 105, 107
  as source of distress, 106
  as stimulus to travel, 103, 104, 107, 116

urban ecology, 2, 3, 5, 7, 11
'urban penalty', 12, 15, 19, 21
urbanisation, 22, 81, 90, 100, 101

vagrancy, 94, 100
Valentine, James, 94, 99
Vance, James, 2
Vienna, 91
voluntary principle, 101

Wakefield, 56
wakes, 35, 44, 45
Ward, David, 3, 4, 8
Watkins, Miles, 107, 117
Watson, William, 93, 94
Weber, Max, 2

Welsh, 57, 77
White, Jerry, 34
Whitechapel, 101
widows, 20
Williams, Anthony, 23, 26
Williams, Raymond, 7, 80, 81, 82
women, 10, 40, 87, 97, 99, 101
woollen workers, 104–5
workers
    skilled, 10, 15, 32, 33, 43, 88, 104, 114
    unskilled, 10, 32, 41, 114, 121
workhouse, 13, 19
Wright, Thomas, 117, 120

York, 56
Yorkshire, 106, 107, 112, 126, 129